Data Science Certification Program for Game Changers

Learn. Practice. Get Certified.

ANGEL BERNIZ

International Institute of Executive Careers (IIEC)

Program Contents

Welcome to the Program

Thank you very much for purchasing this book.

First of all, let's connect on LinkedIn and join my **Game Changers** Newsletter on LinkedIn, so that you can get my insights, learnings, and exclusive contents to help you get career success:

https://www.linkedin.com/in/angelberniz

I would love you to share your experience with me after completing the program. You can message me via LinkedIn.

Okay, let's move on!

Certification Program

This book is the foundation for the **IIEC Data Science Certification Program™** available here:

https://thebestmba.org/program/data-science-certification

The Certification Program is available in the online program and will improve your job opportunities and certify your skills gained from this book.

The exams in the online program are based on the exam questions covered along this book. And once you sign up, you can take your online program whenever you want (no deadline limits). Just sign up today for your Certification Program and certify your skills.

Now, let's start enjoying this journey.

Preface

As we stand at the dawn of the 21st century, we find ourselves at a pivotal moment in history, where rapid advancements in Data Science and related technologies are dramatically reshaping the way we understand and interact with the world around us. The Data Science era presents both unprecedented challenges and transformative opportunities for those who aspire to lead the change and make a lasting impact. It is increasingly urgent for these game changers to adopt a thoughtful, ethical, and inclusive approach to navigating this complex and evolving landscape.

The purpose of this program is to provide a practical and accessible resource for game changers, empowering them to harness the potential of Data Science and lead the way in shaping the future of data-driven decision-making. Drawing on insights from a diverse array of disciplines, including computer science, statistics, social sciences, humanities, and ethics, this program seeks to foster a holistic and interdisciplinary understanding of the Data Science era and its implications for various sectors.

Throughout the program, we will lay a foundation for understanding the key principles and concepts that underpin Data Science technologies and discuss the ethical considerations that arise as we integrate these advancements into various industries. The program offers strategies, best practices, and actionable insights designed to empower game changers in their pursuit of innovation and positive change.

As we delve deeper into the future of data-driven decision-making, we will examine the potential impact of Data Science technologies on employment, education, healthcare, and other sectors, highlighting both the transformative opportunities and the challenges that lie ahead.

Throughout the program, we will underscore the need for responsible innovation, interdisciplinary collaboration, public engagement, and global cooperation to navigate the challenges and seize the opportunities presented by the Data Science era. By taking an active role in shaping the future of data-driven decision-making, game changers can ensure that Data Science technologies are developed and deployed in ways that align with our collective values, ethical principles, and societal goals.

In offering this program, it is my hope that participants will come away with a deeper understanding of the Data Science era and its implications for various sectors, as well as a sense of optimism, inspiration, and empowerment. As game changers, you have a unique opportunity to shape the course of Data Science development in ways that reflect our shared aspirations for a more just, equitable, and sustainable world. Together, let us embrace the challenges and opportunities of the Data Science era and chart a bright future for humanity and data-driven insights, hand in hand.

Acknowledgments

My notes at the university were: "Early adopters are the wealthiest people on earth."

Little did I know how those words would shape my path. As I reflect on my journey, I am overwhelmed with gratitude for the countless individuals who accompanied, guided, and inspired me along the way.

I owe a profound debt of gratitude to my mentors, colleagues, and collaborators in executive education and the technology industry. Their unwavering support and willingness to share their knowledge and perspectives have been invaluable. Their guidance has shaped my career and fueled my passion.

To my students at the International Institute of Executive Careers (IIEC), you have been my constant inspiration. Your curiosity and dedication have made me a better educator and challenged me to refine my ideas. Your insights have significantly contributed to this book.

I also acknowledge the remarkable researchers, professionals, and practitioners who have paved the way for our understanding of the potential of emerging technologies. Their visionary thinking and tireless pursuit of knowledge continue to inspire individuals worldwide, including myself.

I am deeply grateful for the unwavering love, support, and understanding of my friends and family. Your belief in me and willingness to engage with my ideas have given me strength and resilience. This book is dedicated to you with love and gratitude.

As I conclude, I am reminded of the interconnected relationships, experiences, and ideas that have shaped this book. To all who have touched my life, big or small, thank you. Your presence has been truly meaningful. Together, let us continue exploring the evolving landscape of technology, guided by the belief that early adopters can shape the future.

About the Author

Angel Berniz is a seasoned Business and Technology executive, strategist, and advisor with over 20 years of experience in the technology industry. He has held various leadership positions in several companies and has worked with clients in different industries, helping them navigate cloud transformation, innovation, sustainability, emerging technologies, and cybersecurity.

Angel Berniz is also a recognized keynote speaker and author on topics related to business technologies, leadership, and innovation. He has been featured in several conferences, seminars, and publications. He is also a professor and lecturer in international executive programs of several Business Schools.

You can explore Angel Berniz's extensive collection of books available on Amazon:

https://www.amazon.com/author/angelberniz

Module 1. Introduction to Data Science

Introduction to Data Science

Welcome to Module 1 on What is Data Science. In this module, you will learn about the exciting field of data science, its importance in the modern world, and how it has revolutionized the way businesses, governments, and researchers tackle complex problems.

Data Science is a multidisciplinary field that combines aspects of statistics, computer science, and domain expertise to extract valuable insights from large amounts of structured and unstructured data. It involves the application of scientific and analytical methods to explore, analyze, visualize, and interpret data to make informed decisions, optimize processes, and predict future trends.

The exponential growth of data, often referred to as "Big Data," has created a significant demand for professionals who can harness its potential. The ability to understand and interpret data is becoming increasingly vital across various industries, including healthcare, finance, marketing, and technology, to name a few. Data scientists use their skills to uncover hidden patterns, identify correlations, and provide valuable insights that can drive strategic decision-making.

For instance, in the healthcare industry, data science can help predict disease outbreaks, optimize patient care, and identify new treatment methods. In the financial sector, data scientists use algorithms and models to detect fraud, assess risks, and make more informed investment decisions. In marketing, data science is applied to analyze customer behavior, segment markets, and personalize content to improve customer engagement and increase revenue.

In this module, you will learn about the fascinating world of data science and the impact it has on various industries. We will introduce you to the key concepts and techniques, and explore how data science is transforming the way we live and work. Get ready to embark on a journey into the world of data science, where the possibilities are endless and the potential for innovation is limitless.

As you progress through this course, you will develop a strong foundation in data science, acquire essential skills, and gain practical experience in solving real-world problems. By the end of this journey, you will be well-equipped to excel in this rapidly growing and highly sought-after field.

What is Data Science?

Data Science is an interdisciplinary field that combines aspects of statistics, computer science, and domain expertise to derive valuable insights and knowledge from vast amounts of data. At its core, data science aims to make sense of both structured and unstructured data, using analytical, statistical, and programming skills to solve complex problems and drive strategic decision-making.

As a data scientist, you will be responsible for collecting, cleaning, and analyzing data to uncover hidden patterns and trends, identify correlations, and generate actionable insights. To accomplish this, you will use various tools, algorithms, and machine learning techniques to process and interpret data, ultimately transforming raw data into valuable information.

Let's explore some practical examples to better understand the role of data science in everyday life:

- **Personalized Recommendations:** As you browse your favorite e-commerce website, you may notice product recommendations tailored to your interests and preferences. Data science techniques are used to analyze your browsing history, past purchases, and other users' behavior to provide you with personalized suggestions that increase the likelihood of making a purchase.

- **Fraud Detection:** When your credit card company alerts you to a suspicious transaction, it is the result of data science algorithms working behind the scenes. By analyzing patterns in your transaction history and comparing them to known fraudulent activities, data science helps identify potential threats to your financial security.

- **Traffic Management:** Data science is used to analyze and predict traffic patterns, enabling you to find the most efficient route to your destination. By collecting data from various sources, such as GPS devices, traffic cameras, and social media, data science techniques can provide real-time traffic updates and optimize your travel time.

- **Healthcare:** Data science plays a significant role in the healthcare sector by helping to predict disease outbreaks, optimize patient care, and develop new treatment methods. By analyzing electronic health records, genetic data, and other relevant information, data science can identify patterns and correlations that lead to improved patient outcomes.

As a data scientist, you will have the unique opportunity to apply your skills and knowledge to solve real-world problems and contribute to the development of innovative solutions across a wide range of industries. By mastering data science techniques, you will become an invaluable asset to any organization, driving growth, efficiency, and success in the age of Big Data.

The Data Science Process

The Data Science Process is a series of steps that guide you through the various stages of a data science project, from understanding the problem to delivering actionable insights. By following this structured approach, you can ensure that your project is well-organized, efficient, and ultimately successful. Here, we will outline the main steps of the Data Science Process and provide practical examples to help you understand each stage in detail.

1. **Define the Problem:** The first step in the Data Science Process is to identify and understand the problem you want to solve. This involves discussing the objectives and requirements with stakeholders, understanding the desired outcomes, and framing the problem in a data-driven context. For example, you might want to predict customer churn in a subscription-based business to improve retention rates and increase revenue.

2. **Data Collection:** Once you have defined the problem, the next step is to collect the necessary data. This may involve gathering data from various sources such as databases, APIs, web scraping, or third-party providers. For instance, to predict customer churn, you would need data on customer demographics, subscription history, usage patterns, and other relevant information.

3. **Data Cleaning and Preprocessing:** After collecting the data, you will need to clean and preprocess it to ensure that it is accurate, consistent, and suitable for analysis. This stage involves handling missing values, correcting inconsistencies, removing duplicates, and transforming data into the appropriate format. For example, you might standardize dates, remove outliers, or encode categorical variables as numerical values.

4. **Data Exploration and Visualization:** In this stage, you will explore and visualize the data to gain a deeper understanding of its structure, relationships, and patterns. This involves using descriptive statistics, plotting graphs, and creating visual representations of the data. Through this process, you might discover correlations between customer usage patterns and their likelihood to churn, which can help inform your predictive model.

5. **Feature Engineering and Selection:** Feature engineering involves creating new features from the existing data to improve the performance of your predictive model. Feature selection involves selecting the most relevant features that contribute to the prediction. For instance, you might create a new feature that calculates the average time between customer logins, or you might remove redundant features that have little impact on the prediction.

6. **Model Building and Evaluation:** In this step, you will build and evaluate machine learning models to make predictions based on the data. This involves selecting an appropriate algorithm, training the model on your dataset, and evaluating its performance using metrics such as accuracy, precision, recall, or F1-score. For the customer churn example, you might compare various algorithms like logistic regression, decision trees, or support vector machines to find the best-performing model.

7. **Model Deployment and Monitoring:** After selecting the best model, you will deploy it in a production environment to make predictions on new data. This might involve integrating the model into an existing system, creating an API, or developing a user interface. Once the model is deployed, you will need to monitor its performance and update it as needed to ensure that it remains accurate and relevant.

Following the Data Science Process will allow you to systematically approach any data-driven problem, increasing the likelihood of success and delivering valuable insights to inform decision-making. As you gain experience and refine your skills, you will become increasingly adept at navigating each stage of the process, making you a highly effective data scientist.

Now let's dive into the details of each of these data science process steps:

Step 1: Define the Problem

Defining the problem is a critical first step in the data science process, as it sets the foundation for the entire project. A well-defined problem allows you to focus your efforts and ensure that the project's goals are aligned with the stakeholders' expectations. In this stage, you should aim to develop a clear understanding of the objectives, identify the relevant metrics for evaluation, and establish the project's scope and constraints.

Here are some key aspects to consider when defining the problem:

- **Understand the business context:** Before diving into data analysis, it's crucial to understand the business context and the industry's domain-specific knowledge. This helps you to identify the right questions to ask and to recognize the potential value that data science can bring to the organization. For example, if you're working for a manufacturing company, you should familiarize yourself with the production processes, the challenges they face, and their key performance indicators (KPIs).

- **Identify the stakeholders:** Determine who the stakeholders are and understand their expectations and requirements. Stakeholders can include business executives, managers, customers, or other team members. Engaging with stakeholders helps you to understand their perspectives, identify potential roadblocks, and establish a shared vision for the project's success.

- **Set clear objectives:** Establish clear and measurable objectives for the project. These objectives should be Specific, Measurable, Achievable, Relevant, and Time-bound (SMART). For example, instead of setting a vague goal like "improve customer satisfaction," you could aim to "increase the Net Promoter Score (NPS) by 10% within six months."
- **Determine the evaluation metrics:** Decide on the appropriate metrics to measure the success of your project. These metrics will help you assess the performance of your data science models and quantify the value of your work. For instance, if your goal is to increase customer retention, you might use metrics such as churn rate, customer lifetime value (CLV), or customer acquisition cost (CAC).
- **Identify data sources and availability:** Consider the data sources that you will need to address the problem and assess their availability. This includes evaluating the quality, format, and accessibility of the data. You might need to collect data from internal databases, external APIs, or third-party data providers. For example, if you're working on a project to optimize marketing campaigns, you might need data on customer demographics, past campaign performance, and online behavior.
- **Establish project scope and constraints:** Determine the scope of your project, including the timeline, resources, and budget. Establishing the project's boundaries helps you manage expectations and ensure that the project remains focused and feasible. For example, you might need to prioritize certain tasks, limit the scope of the analysis, or work within a tight budget.

Defining the problem will allow you to create a solid foundation for your data science project, ensuring that your efforts are targeted, relevant, and valuable. This will help you stay focused throughout the data science process, increase the likelihood of success, and deliver meaningful results that align with your organization's goals.

Step 2: Data Collection

Data collection is a crucial step in the data science process, as the quality and relevance of the data you gather will directly impact the success of your project. In this stage, you aim to acquire the necessary data to address the problem you've defined, ensuring that the data is accurate, comprehensive, and representative. Here are some key aspects to consider when collecting data for your project:

- **Data identification:** Start by identifying the data sources that are relevant to your project. These sources may include internal databases, external APIs, web scraping, surveys, or third-party data providers. For example, if you're working on a project to predict house prices, you might need data on property features, location, historical prices, and market trends.

- **Data access and extraction:** Once you've identified the data sources, you need to access and extract the data. This may involve requesting access to internal databases, using APIs to query external services, or writing web scraping scripts to extract data from websites. When accessing data, be sure to follow the relevant legal and ethical guidelines, and obtain any necessary permissions.

- **Data storage:** Determine the appropriate storage solutions for your data, considering factors such as scalability, performance, and security. You might store the data in local files, relational databases, or distributed storage systems, depending on the size and complexity of the dataset. For example, you could store a small dataset in a

CSV file or a large dataset in a distributed database like Hadoop or Apache Cassandra.

- **Data format and structure:** Consider the format and structure of the data you collect. Data can be structured (e.g., numerical, categorical) or unstructured (e.g., text, images, audio). It's essential to understand the structure of your data, as this will determine the preprocessing techniques and analytical methods you'll need to use. For instance, if you're working with text data, you might need to employ natural language processing (NLP) techniques to extract meaningful information.

- **Data quality assessment:** Assess the quality of the data you collect, looking for issues such as missing values, inconsistencies, outliers, or errors. Identifying and addressing data quality issues early in the process will help ensure that your analysis is accurate and reliable. For example, you might need to clean or impute missing values, correct inconsistencies, or remove duplicates before proceeding with your analysis.

- **Data integration:** In many cases, you'll need to combine data from multiple sources to create a comprehensive dataset for your project. Data integration involves merging and consolidating data from different sources, ensuring that the data is consistent and compatible. This may require you to transform data formats, align data schemas, or resolve conflicts in the data. For instance, if you're working with data from multiple marketing channels, you might need to standardize the naming conventions and units of measurement across the sources.

Collecting and curating the data for your project will allow you to set the stage for successful analysis and modeling. High-quality data is the foundation of effective data science, enabling you to uncover meaningful insights, build accurate models, and deliver valuable results that drive decision-making and innovation.

Step 3: Data Cleaning and Preprocessing

Data cleaning and preprocessing are essential steps in the data science process, ensuring that your data is accurate, consistent, and suitable for analysis. By addressing issues such as missing values, outliers, and inconsistencies, you can improve the quality of your dataset and increase the reliability of your results. Here are some key aspects to consider when cleaning and preprocessing your data:

- **Handling missing values:** Missing values can occur for various reasons, such as data entry errors, system limitations, or incomplete records. To address missing values, you can either remove the affected records, fill in the missing data using an appropriate imputation method, or incorporate the missingness as a feature in your analysis. For example, you might use the mean or median to impute missing numerical values or the mode for missing categorical values.

- **Removing duplicates:** Duplicate records can distort your analysis and lead to inaccurate results. To address this issue, you should identify and remove any duplicates from your dataset. This might involve comparing records based on unique identifiers, such as customer IDs or transaction numbers, or using a combination of features to identify potential duplicates.

- **Correcting inconsistencies:** Inconsistencies in your data can arise from various sources, such as data entry errors, variations in units of measurement, or discrepancies between data sources. To address inconsistencies, you should standardize your data by applying consistent naming conventions, units of measurement, and data formats. For example, you might need to convert temperatures from Fahrenheit to Celsius or standardize date formats across your dataset.

- **Encoding categorical variables:** Many machine learning algorithms require that input features be represented as numerical values. To accommodate this requirement, you'll need to encode categorical variables as numerical values. Common encoding techniques include one-hot encoding, which creates binary features for each category, or ordinal encoding, which assigns an integer value to each category based on a predefined order.

- **Scaling and normalization:** Machine learning algorithms can be sensitive to the scale of input features, which can affect their performance and convergence. To address this issue, you can scale and normalize your data, ensuring that all features are on a comparable scale. Common scaling techniques include min-max scaling, which scales features to a specific range, and standard scaling, which centers features around the mean and scales them based on the standard deviation.

- **Feature transformation:** In some cases, you might need to transform your features to better represent the underlying patterns in the data or to meet the assumptions of a particular algorithm. Common feature transformations include log transformations, which can reduce the impact of outliers or skewed distributions, and power transformations, which can help stabilize the variance across features.

- **Outlier detection and treatment:** Outliers are data points that deviate significantly from the overall pattern in the data. Outliers can arise from data entry errors, measurement errors, or genuine variations in the data. To address outliers, you can either remove them from your dataset, cap them at a predetermined threshold, or use robust methods that are less sensitive to outliers in your analysis.

Cleaning and preprocessing your data will allow you to ensure that your analysis is based on accurate, consistent, and representative data. This will help you uncover meaningful insights, build robust models, and deliver reliable results that drive decision-making and innovation in your organization.

Step 4: Data Exploration and Visualization

Data exploration and visualization are critical steps in the data science process, allowing you to gain a deeper understanding of your dataset's structure, relationships, and patterns. By exploring your data and creating visual representations, you can identify trends, detect anomalies, and generate hypotheses to guide your analysis. Here are some key aspects to consider when exploring and visualizing your data:

- **Descriptive statistics:** Start your data exploration by calculating descriptive statistics, such as the mean, median, mode, standard deviation, and percentiles. These summary statistics can provide insights into the central tendency, dispersion, and shape of your data's distribution. For example, you might calculate the average revenue per customer or the standard deviation of product prices.

- **Correlation analysis:** Investigate the relationships between pairs of variables to identify potential associations or dependencies. Correlation analysis can help you detect linear relationships between variables, which can inform your feature selection and modeling decisions. For example, you might discover that customer satisfaction is positively correlated with repeat purchases, indicating a potential area for further investigation.

- **Univariate analysis:** Examine the distribution and characteristics of individual variables in your dataset. This can involve plotting histograms, box plots, or kernel density estimations to visualize the distribution of continuous variables, or bar charts and pie charts to represent categorical variables. For example, you might create a histogram to visualize the distribution of customer ages or a bar chart to display the frequency of different product categories.

- **Bivariate and multivariate analysis:** Explore the relationships between two or more variables by creating scatter plots, heatmaps, or parallel coordinate plots. These visualizations can help you identify patterns and trends in the data, such as clusters, outliers, or interactions between variables. For example, you might create a scatter plot to visualize the relationship between product price and customer ratings or a heatmap to display the correlation matrix between multiple features.

- **Dimensionality reduction:** When working with high-dimensional data, it can be challenging to visualize and explore the relationships between variables. Dimensionality reduction techniques, such as Principal Component Analysis (PCA) or t-Distributed Stochastic Neighbor Embedding (t-SNE), can help you project high-dimensional data onto a lower-dimensional space, enabling more effective visualization and exploration. For example, you might use PCA to visualize the relationships between a large set of customer attributes in a two-dimensional plot.

- **Time series analysis:** If your data includes a temporal component, consider performing time series analysis to explore trends, seasonality, and patterns over time. This can involve creating line charts, autocorrelation plots, or decompositions to visualize the various components of your time series data. For example, you might create a line chart to visualize the monthly sales revenue or an autocorrelation plot to examine the seasonality in customer visits.

- **Interactive visualizations:** Create interactive visualizations to enable more in-depth exploration and analysis of your data. Interactive visualizations allow users to manipulate the view, filter data, or drill down into specific details. Tools such as Plotly, Bokeh, or Tableau can help you create dynamic and engaging visualizations for your data. For example, you might create an interactive dashboard to monitor the performance of marketing campaigns in real-time.

Exploring and visualizing your data will allow you to gain valuable insights, identify potential issues, and guide your analysis and modeling decisions. This will help you uncover meaningful patterns, build robust models, and deliver impactful results that drive decision-making and innovation in your organization.

Step 5: Feature Engineering and Selection

Feature engineering and selection are essential steps in the data science process, as they help you transform and select the most relevant variables to use in your analysis and modeling. By creating new features and choosing the most informative ones, you can improve the performance, interpretability, and generalizability of your models. Here are some key aspects to consider when performing feature engineering and selection:

- **Feature creation:** Generate new features by combining, transforming, or aggregating existing variables in your dataset. This can involve mathematical operations, domain-specific transformations, or the extraction of information from complex data types, such as text, images, or time series. For example, you might calculate the ratio of income to expenses for a financial dataset or extract sentiment scores from customer reviews using natural language processing techniques.

- **Feature transformation:** Apply transformations to your features to improve their distribution, scale, or relationship with the target variable. Common transformations include log, square root, or power transformations, which can help stabilize variance, reduce the impact of outliers, or linearize relationships. For example, you might apply a log transformation to a variable with a skewed distribution, such as income or population size.

- **Feature scaling:** Scale your features to ensure that they have comparable magnitudes and are suitable for use with machine learning algorithms that are sensitive to feature scales. Common scaling methods include min-max scaling, which scales features to a specific range, and standard scaling, which centers features around the mean and scales them based on the standard deviation. For example, you might apply standard scaling to a dataset containing features with different units of measurement, such as age in years and income in dollars.

- **Feature encoding:** Encode categorical variables as numerical values to make them compatible with machine learning algorithms that require numerical inputs. Common encoding techniques include one-hot encoding, which creates binary features for each category, and ordinal encoding, which assigns an integer value to each category based on a predefined order. For example, you might apply one-hot encoding to a categorical variable representing customer segments or ordinal encoding to a variable representing education levels.

- **Feature selection:** Identify the most relevant and informative features to include in your analysis and modeling. Feature selection techniques can help you reduce the dimensionality of your dataset, improve model performance, and enhance interpretability. Common feature selection methods include filter methods, which rank features based on their relationship with the target variable, and wrapper methods, which evaluate subsets of features using a specific algorithm. For example, you might use correlation analysis to identify highly correlated features or employ recursive feature elimination to find the optimal subset of features for a particular model.

- **Feature importance:** Assess the importance or contribution of individual features to the performance of your model. This can involve techniques such as permutation importance, which measures the change in model performance when a feature's values are randomly shuffled, or model-specific importance metrics, such as Gini importance for decision trees or coefficient magnitudes for linear models. Understanding feature importance can help you interpret your model's results, identify potential issues, and guide further feature engineering efforts.

Engineering and selecting features will allow you to ensure that your analysis and modeling are based on the most relevant and informative variables. This will help you build more accurate, interpretable, and generalizable models, ultimately leading to more impactful results that drive decision-making and innovation in your organization.

Step 6: Model Building and Evaluation

Model building and evaluation are critical steps in the data science process, as they involve selecting, training, and assessing machine learning algorithms to find the best model for your specific problem. By carefully building and evaluating models, you can identify the most effective and interpretable solutions, ultimately delivering reliable and actionable results. Here are some key aspects to consider when building and evaluating models:

- **Algorithm selection:** Choose appropriate machine learning algorithms based on your problem type, data characteristics, and modeling goals. Consider factors such as interpretability, complexity, and scalability when selecting an algorithm. For example, you might choose linear regression for a simple and interpretable regression problem or a deep learning model for a complex image classification task.

- **Data splitting:** Divide your dataset into separate training, validation, and testing sets. This allows you to train your model on one set of data and evaluate its performance on unseen data, ensuring that your model is generalizable and not overfitting. A common approach is to use a 70-15-15 or 80-10-10 split for the training, validation, and testing sets, respectively.

- **Feature selection and preprocessing:** Apply feature selection, scaling, encoding, and other preprocessing techniques to your data as needed, ensuring that your input features are compatible with the chosen algorithm and have the potential to improve model performance. For example, you might perform dimensionality reduction using Principal Component Analysis (PCA) or standardize your features using z-score scaling.

- **Model training:** Train your model on the training dataset, adjusting its parameters and learning from the data. This may involve setting initial hyperparameters, selecting a loss function, and choosing an optimization algorithm. Be mindful of the trade-off between model complexity and interpretability, as well as the risk of overfitting or underfitting.

- **Model validation and tuning:** Evaluate your model's performance on the validation dataset, using appropriate performance metrics based on your problem type, such as accuracy, precision, recall, F1 score, or mean squared error. Adjust the model's hyperparameters and retrain the model as needed to optimize its performance. Techniques like grid search, random search, or Bayesian optimization can help you systematically explore the hyperparameter space and identify the best configuration.

- **Model testing and evaluation:** Once you have fine-tuned your model, assess its performance on the testing dataset, which has not been used in the training or validation process. This provides an unbiased estimate of your model's performance and generalizability. Compare the performance of different models or algorithms to identify the best solution for your problem.

- **Model interpretability and explainability:** Ensure that your model is interpretable and can provide meaningful explanations for its predictions. This can involve techniques such as feature importance analysis, partial dependence plots, or model-specific explanation methods, such as LIME or SHAP for complex models. An interpretable model can enhance trust, facilitate understanding, and aid in decision-making.
- **Model validation with real-world data:** If possible, validate your model's performance using real-world data that reflects the conditions under which it will be deployed. This can help you identify potential issues and ensure that your model is robust and effective in practice.

Building and evaluating models will allow you to find the most effective and interpretable solutions for your specific problem, delivering reliable and actionable results that drive decision-making and innovation in your organization.

Step 7: Model Deployment and Monitoring

Model deployment and monitoring are crucial aspects of the data science process that enable you to use your trained machine learning models in real-world applications and track their performance over time. By effectively deploying and monitoring your models, you can provide valuable insights and predictions that support decision-making and innovation in your organization. Here are some key aspects to consider when deploying and monitoring models:

- **Deployment options:** Determine the most suitable deployment option for your model, considering factors such as latency, cost, scalability, and security. Some popular deployment options include:

- **RESTful APIs:** Deploy your model as a RESTful API to enable easy integration with web applications, mobile apps, or other systems. Tools like Flask, FastAPI, or Django can help you build and deploy RESTful APIs for your models.
- **Containerization:** Containerize your model using technologies like Docker or Kubernetes to ensure consistent behavior, easy scaling, and seamless deployment across different environments and platforms.
- **Serverless functions:** Utilize serverless platforms like AWS Lambda, Google Cloud Functions, or Azure Functions to deploy your model as a scalable, pay-per-use service, without the need to manage infrastructure.

- **Model monitoring:** Implement a monitoring system to track your model's performance and usage in real-time. Key aspects of model monitoring include:
  - **Performance metrics:** Continuously monitor performance metrics such as accuracy, precision, recall, F1 score, or mean squared error to ensure your model is maintaining its effectiveness and meeting your organization's requirements.
  - **Resource utilization:** Track resource usage, such as CPU, memory, and storage, to ensure your model is running efficiently and identify potential bottlenecks or capacity issues.
  - **Data quality and drift:** Monitor the quality and distribution of incoming data to detect any changes that may impact your model's performance, such as

data drift or concept drift. Tools like TensorFlow Data Validation or Great Expectations can help you validate and monitor data quality.

- **Alerting and notifications:** Set up alerting and notification systems to inform stakeholders about critical events or changes in your model's performance. This can help you proactively address issues and maintain the reliability and effectiveness of your model.

- **Model validation:** Periodically validate your model using real-world data to ensure it remains accurate and relevant in the context of your application. This can involve collecting new data, updating training and testing sets, and re-evaluating your model's performance.

- **A/B testing:** Conduct A/B testing or multi-armed bandit experiments to compare different model versions or algorithms and identify the most effective solution for your problem. This can help you make data-driven decisions about model updates and enhancements.

- **Model governance:** Establish a model governance framework to ensure your models are developed, deployed, and maintained in accordance with your organization's policies, regulations, and ethical guidelines. This can involve implementing best practices for data privacy, security, and compliance, as well as maintaining thorough documentation and audit trails.

Deploying and monitoring your models will allow you to deliver actionable insights and predictions that support decision-making and innovation in your organization while ensuring their continued effectiveness, reliability, and compliance with relevant policies and regulations.

The Role of a Data Scientist

As a data scientist, your role is to extract valuable insights and knowledge from data, using a combination of domain expertise, programming skills, and statistical and machine learning techniques. Your work can significantly impact decision-making and drive innovation within your organization. Here are some key responsibilities and tasks that you may undertake as a data scientist:

- **Problem identification:** Work closely with stakeholders to understand their needs, identify problems that can be addressed using data, and translate these problems into data-driven questions or objectives. This requires strong communication and collaboration skills, as well as the ability to think critically and analytically.

- **Data acquisition and preparation:** Collect and preprocess data from various sources, such as databases, APIs, or external platforms, ensuring its quality, relevance, and integrity. This involves tasks like data cleaning, transformation, and encoding, and requires proficiency in programming languages like Python or R, as well as knowledge of SQL and data manipulation libraries.

- **Exploratory data analysis:** Perform exploratory data analysis (EDA) to gain a better understanding of your data, identify patterns and relationships, and uncover potential issues or anomalies. This can involve techniques like summary statistics, data visualization, and correlation analysis, and requires a solid foundation in statistics and data visualization tools.

- **Feature engineering and selection:** Create new features and select the most informative ones for your analysis and modeling. This involves leveraging domain knowledge, statistical techniques, and machine learning algorithms to transform, aggregate, or encode variables, ultimately enhancing the performance and interpretability of your models.

- **Model building and evaluation:** Select, train, and evaluate machine learning models to address the identified problem, using appropriate algorithms, performance metrics, and validation techniques. This requires a deep understanding of machine learning algorithms, their assumptions, and trade-offs, as well as the ability to implement and fine-tune these models using programming libraries and frameworks.

- **Model deployment and monitoring:** Deploy your trained models into production environments and set up monitoring systems to track their performance and usage over time. This involves working with cloud platforms, APIs, containerization technologies, and monitoring tools, ensuring that your models are reliable, scalable, and actionable.

- **Communication and visualization:** Communicate your findings, insights, and recommendations to stakeholders using clear language, effective visualizations, and storytelling techniques. This requires strong presentation and data visualization skills, as well as the ability to tailor your message to different audiences and contexts.

- **Collaboration and teamwork:** Work effectively within interdisciplinary teams, collaborating with data engineers, business analysts, domain experts, and other professionals to ensure the success of your data-driven projects. This requires strong interpersonal skills, adaptability, and the ability to learn and apply new concepts and techniques.

- **Continuous learning:** Stay up-to-date with the latest developments in data science, machine learning, and your domain, continually expanding your knowledge and skills.

This involves reading research papers, attending conferences and workshops, and participating in online forums or communities.

Mastering these responsibilities and tasks will allow you to excel as a data scientist and make a significant impact within your organization, driving data-driven decision-making and innovation that lead to better outcomes and increased value.

Key Skills for Data Scientists

To succeed as a data scientist, you need to develop a diverse set of skills that span multiple disciplines. Acquiring these skills will enable you to tackle complex data-driven problems and deliver valuable insights and predictions to your organization. Here are some key skills that you should focus on as a data scientist:

1. **Programming:** Ensure data scientists have proficiency in programming languages like Python or R, which are essential for data manipulation, analysis, and model building. This knowledge enables them to work with data science libraries and implement advanced techniques more efficiently.

2. **Statistics:** A strong foundation in statistics is vital for understanding underlying patterns and relationships in data and making informed decisions about modeling and evaluation. Data scientists should be familiar with various statistical concepts and techniques.

3. **Machine learning:** Data scientists need to comprehend various machine learning algorithms, their assumptions, and trade-offs. This understanding allows them to select appropriate models and techniques for different problems, leading to more accurate and effective solutions.

4. **Data manipulation and preprocessing:** Expertise in cleaning, transforming, and preprocessing data is crucial for ensuring data quality and relevance for analysis. Data scientists should be skilled in various data preparation techniques, enabling them to feed accurate and useful data into their models.

5. **Data visualization:** Data visualization skills are important for exploring data effectively and communicating findings to stakeholders. Data scientists should be proficient in

creating clear and compelling visualizations that highlight key insights and patterns in the data.

6. **Domain expertise:** Data scientists should possess a deep understanding of their domain, allowing them to apply data science skills more effectively and generate valuable insights for the organization. This may involve learning about specific industries, business processes, or subject areas.

7. **Big data technologies:** Familiarity with big data technologies and platforms, such as Hadoop, Spark, and NoSQL databases, can help data scientists process, store, and analyze large-scale datasets. This knowledge is valuable when working with massive datasets or in organizations relying heavily on big data infrastructure.

8. **Cloud computing:** Understanding cloud computing platforms, such as AWS, Google Cloud, or Azure, can help data scientists deploy and manage their models more efficiently and at scale. This skill is particularly important in organizations that leverage cloud services for data storage, processing, and machine learning.

9. **Communication and presentation:** Strong communication and presentation skills are crucial for data scientists to effectively convey their findings, insights, and recommendations to stakeholders. They should be able to create engaging and persuasive presentations tailored to different audiences and contexts.

10. **Collaboration and teamwork:** Data scientists must have interpersonal skills and the ability to work effectively within interdisciplinary teams. They should be able to collaborate with other professionals, such as data engineers and business analysts, to ensure the success of data-driven projects and adapt to new concepts and techniques as needed.

Understanding the importance of these key skills and fostering their development within your data science team will allow you to drive data-driven decision-making and innovation within your organization, leading to better outcomes and increased value.

Here you can find how these skills are covered in this program:

1. **Programming:**
   - Module 2: Introduction to Python for Data Science
   - Module 3: Python Libraries for Data Science
2. **Statistics:**
   - Module 4: Foundations of Statistics for Data Science
3. **Machine learning:**
   - Module 5: Supervised Machine Learning
   - Module 6: Unsupervised Machine Learning
   - Module 7: Deep Learning and Reinforcement Learning
4. **Data manipulation and preprocessing:**
   - Module 8: Data Wrangling and Preprocessing
5. **Data visualization:**
   - Module 9: Data Visualization Techniques
6. **Domain expertise:**
   - Throughout the book, you will find examples and case studies from various domains, which will help you understand the applications of data science techniques in different industries. You are encouraged to apply the knowledge and techniques

learned from the book to your specific area of interest.

7. **Communication and presentation:**
    - Module 9: Data Visualization Techniques (includes aspects of communication and presentation)

While the book may touch upon **big data technologies**, **cloud computing**, and **collaboration and teamwork**, these topics may not be covered in-depth in the provided Table of Contents. You can use the foundational knowledge gained from the book to explore these areas further through additional resources or programs provided by the IIEC at https://thebestmba.org

Module 1 Exam Questions

1. Which programming language is commonly used for data manipulation and analysis in data science?

a) Java

b) Python

c) C++

d) JavaScript

2. Which of the following is a key component of descriptive statistics?

a) Mean

b) Standard Deviation

c) Median

d) All of the above

3. What is the primary purpose of data cleaning in the data science process?

a) Identifying patterns in data

b) Visualizing data

c) Ensuring data quality and relevance

d) Building predictive models

4. What is the main goal of supervised machine learning?

a) Identifying clusters within data

b) Reducing the dimensionality of data

c) Predicting the target variable based on input features

d) Learning patterns without any guidance

5. Which of the following is a popular Python library for data manipulation and analysis?

a) TensorFlow

b) NumPy

c) Pandas

d) Scikit-learn

6. Which clustering algorithm groups data points based on their density in the feature space?

a) K-Means

b) Agglomerative Clustering

c) DBSCAN

d) Spectral Clustering

7. In a neural network, what is the primary purpose of an activation function?

a) To initialize weights and biases

b) To introduce non-linearity into the network

c) To define the network architecture

d) To train the network using backpropagation

8. What is the main purpose of a box plot in data visualization?

a) To show the distribution of a continuous variable

b) To display the relationship between two continuous variables

c) To visualize the frequency of categorical variables

d) To represent the progression of a continuous variable over time

9. Which of the following is a popular dimensionality reduction technique?

a) Principal Component Analysis (PCA)

b) K-Means Clustering

c) Linear Regression

d) Decision Trees

10. What is the primary goal of exploratory data analysis (EDA)?

a) To build predictive models

b) To understand the underlying structure and patterns in the data

c) To deploy machine learning models

d) To monitor the performance of deployed models

11. In which machine learning category does the k-Nearest Neighbors (k-NN) algorithm belong?

a) Supervised learning

b) Unsupervised learning

c) Deep learning

d) Reinforcement learning

12. Which of the following is NOT a key skill for a data scientist?

a) Programming

b) Graphic design

c) Statistics

d) Machine learning

13. What is the main goal of data preprocessing?

a) Building predictive models

b) Preparing data for analysis and modeling

c) Visualizing data

d) Deploying models

14. Which of the following is a popular data visualization library in Python?

a) Matplotlib

b) NumPy

c) Pandas

d) TensorFlow

15. Which of the following best describes the purpose of feature engineering?

a) Creating and selecting the most relevant features for modeling

b) Visualizing the features in a dataset

c) Deploying machine learning models

d) Monitoring the performance of models

16. What does the term "big data" refer to?

a) Large volumes of structured data

b) Large volumes of unstructured data

c) High-velocity data streams

d) All of the above

17. Which of the following algorithms is commonly used for solving classification problems?

a) Linear Regression

b) K-Means Clustering

c) Support Vector Machines (SVM)

d) Principal Component Analysis (PCA)

18. Which of the following is an example of unsupervised machine learning?

a) Sentiment analysis

b) Image recognition

c) Anomaly detection

d) Spam filtering

19. In the context of data science, what does the acronym API stand for?

a) Application Programming Interface

b) Advanced Processing Infrastructure

c) Automated Preprocessing Integration

d) Adaptive Prediction Index

20. What is the primary purpose of reinforcement learning algorithms?

a) Learning optimal actions in an environment by interacting with it

b) Analyzing large volumes of text data

c) Predicting continuous target variables

d) Identifying relationships between categorical variables

Answers:

1. b
2. d
3. c
4. c
5. c
6. c
7. b
8. a
9. a
10. b
11. a
12. b
13. b
14. a
15. a
16. d
17. c
18. c
19. a
20. a

Module 2. Python for Data Science

Introduction to Python for Data Science

Welcome to Module 2 on Introduction to Python for Data Science. In this module, you will learn about the powerful programming language Python and its applications in the field of Data Science. Python has become the language of choice for many Data Scientists due to its versatility, ease of use, and extensive ecosystem of libraries specifically tailored to working with data. As we dive into this module, we will explore how Python can be harnessed to efficiently manipulate, analyze, and visualize data, making it an invaluable tool for your Data Science journey.

To set the stage, we will start by discussing the fundamental building blocks of Python, such as variables, data types, and basic operations. This will provide you with a strong foundation to understand and work with more complex Python constructs. You will be introduced to control structures, such as loops and conditionals, which will enable you to create efficient and dynamic code. Furthermore, we will cover the concept of functions, which are essential for organizing and reusing code in a modular fashion. By understanding these core concepts, you will be well-prepared to tackle more advanced topics and utilize Python effectively for Data Science tasks.

We will also touch upon Object-Oriented Programming (OOP) in Python, an essential paradigm that allows you to create and work with custom objects and classes. This will further enhance your ability to write clean, reusable, and maintainable code, a vital skill for any aspiring Data Scientist.

Throughout this module, you will encounter practical examples that illustrate how Python is used to solve real-world Data Science problems. These examples will not only demonstrate the power of Python as a language for Data Science but also help you develop the intuition and skills necessary to implement your own solutions.

By the end of this module, you will have a solid understanding of Python's fundamental concepts and how to apply them in the context of Data Science. Armed with this knowledge, you will be well-prepared to tackle more advanced topics, such as leveraging Python's rich ecosystem of libraries for Data Science, which we will explore in the following module.

Python Basics

In this section, we will introduce you to the essential Python basics that will serve as the foundation for your Data Science journey. We will cover various fundamental concepts such as variables, data types, and basic operations. By understanding these core elements, you will be better equipped to grasp more advanced Python topics and apply them effectively in your Data Science projects.

Running Python code

To run and develop Python code, you can use various tools and environments depending on your preferences and needs. Here are some popular options:

1. **Online Python Code Editors:** Online code editors are web-based tools that allow you to write, test, and share Python code directly from your browser. They are convenient for trying out small pieces of code or sharing code with others. Some popular online Python editors are:

- **Repl.it**: https://repl.it
- **Google Colab**: https://colab.research.google.com
- **Jupyter Notebook via Binder**: https://mybinder.org

To use an online editor, simply navigate to the website, create a new Python file or notebook, and start writing your code. You can run your code directly in the browser and see the output immediately.

2. **Python Integrated Development Environments (IDEs):** Python IDEs provide a comprehensive environment for writing, debugging, and executing Python code. Some popular Python IDEs are:

- **PyCharm**: https://www.jetbrains.com/pycharm

- **Visual Studio Code with the Python extension**: https://code.visualstudio.com
- **JupyterLab**: https://jupyter.org
- **Spyder**: https://www.spyder-ide.org

To use an IDE, download and install the software, then create a new Python file (with a .py extension) and start writing your code. You can run your code directly from the IDE, which will also provide helpful tools for debugging and code completion.

3. **Running Python Locally (Command Line):** You can also run Python code on your local machine using the command line. First, you need to install Python from the official website (https://www.python.org/downloads/). After installation, open a command prompt (Windows) or terminal (macOS, Linux), and type python or python3 to start the Python interpreter. You can then write and execute Python code directly in the interpreter.
To run a Python script, save your code in a .py file and navigate to the file's directory using the command prompt or terminal. Then, execute the script by typing python filename.py or python3 filename.py, replacing filename with the name of your file.

Regardless of the method you choose, you can use the provided examples and explanations from the previous sections to practice and develop your Python skills. Start by writing simple programs and gradually move on to more complex examples as you become more familiar with the language and its concepts.

Variables and Data Types

In Python, a variable is a symbolic name that represents a value. You can think of it as a container that stores information. Python has several built-in data types, including numbers (integers and floats), strings, and booleans. To create a variable, you simply assign a value to a name using the equals sign (=).

For example, let's create a variable to store an integer value:

```
my_age = 25
```

Here, my_age is a variable that holds the integer value 25.

Similarly, you can create variables for other data types:

```
my_name = "John Doe" # A string
temperature = 23.5 # A float
is_sunny = True # A boolean
```

Basic Operations

Python supports various basic operations, such as arithmetic, comparisons, and logical operations, which you can use to manipulate and evaluate data.

Arithmetic operations include addition, subtraction, multiplication, division, and more:

```
a = 10
b = 3
addition = a + b # 13
subtraction = a - b # 7
multiplication = a * b # 30
division = a / b # 3.3333
```

Comparison operators, such as less than (<), greater than (>), equal to (==), and not equal to (!=), allow you to compare values and produce boolean results:

```
a = 10
b = 3
result1 = a < b # False
result2 = a > b # True
result3 = a == b # False
```

```
result4 = a != b # True
```

Logical operators, such as and, or, and not, allow you to combine boolean values:

```
is_sunny = True
is_warm = False
result1 = is_sunny and is_warm # False
result2 = is_sunny or is_warm # True
result3 = not is_sunny # False
```

Mastering these Python basics will allow you to be well-equipped to create and manipulate variables, perform calculations, and evaluate conditions. These foundational skills will prove invaluable as you continue to explore Python's capabilities in Data Science. In the next section, we will delve into control structures and functions, which will enable you to create more complex and dynamic programs.

Data Structures

In this section, you will learn about the various built-in data structures in Python, their properties, use cases, and how to perform common operations with them. Understanding these data structures is essential for handling and manipulating data effectively in your data science projects.

Lists

Lists are ordered, mutable collections of elements. They can store elements of different data types, including other lists. To create a list, use square brackets [] and separate the elements with commas:

```
my_list = [1, 2, 3, "apple", 5.5, [4, 5]]
```

Common list operations include:

- **Accessing elements:** Use the index of the element, starting from 0 for the first element.

```
print(my_list[3]) # Output: apple
```

- **Modifying elements:** Assign a new value to a specific index.

```
my_list[2] = "orange"
```

- **Adding elements:** Use the append() method to add an element to the end of the list, or the insert() method to add an element at a specific position.

```
my_list.append("banana") my_list.insert(1, "grape")
```

- **Removing elements:** Use the remove() method to remove a specific element by value, or the pop() method to remove an element by index.

```
my_list.remove("apple") my_list.pop(2)
```

- **Slicing:** Create a new list by selecting a range of elements from the original list.

```
new_list = my_list[1:4]
```

Tuples

Tuples are ordered, immutable collections of elements. They are similar to lists, but their elements cannot be changed once defined. To create a tuple, use parentheses () and separate the elements with commas:

```
my_tuple = (1, 2, 3, "apple", 5.5)
```

Common tuple operations include:

- **Accessing elements:** Use the index of the element, just like with lists.

```
print(my_tuple[3]) # Output: apple
```

- **Slicing:** Create a new tuple by selecting a range of elements from the original tuple.

```
new_tuple = my_tuple[1:4]
```

Sets

Sets are unordered collections of unique elements. They do not allow duplicate values and do not maintain the order of elements. To create a set, use curly braces {} and separate the elements with commas:

```
my_set = {1, 2, 3, "apple", 5.5}
```

Common set operations include:

- **Adding elements:** Use the add() method to add an element to the set.

```
my_set.add("orange")
```

- **Removing elements:** Use the remove() or discard() method to remove an element from the set.

```
my_set.remove("apple") my_set.discard(2)
```

- **Set operations:** Perform union, intersection, difference, and symmetric difference operations with other sets using methods like union(), intersection(), difference(), and symmetric_difference().

```
set_A = {1, 2, 3} set_B = {2, 3, 4} union_set = set_A.union(set_B)
intersection_set = set_A.intersection(set_B)
```

Dictionaries

Dictionaries are unordered collections of key-value pairs. They are mutable and do not allow duplicate keys. To create a dictionary, use curly braces {} and separate the keys and values with colons, and the key-value pairs with commas:

```
my_dict = {"apple": 1, "orange": 2, "banana": 3}
```

Common dictionary operations include:

- **Accessing elements:** Use the key of the element in square brackets.

```
print(my_dict["apple"]) # Output: 1
```

- **Modifying elements:** Assign a new value to a specific key.

```
my_dict["banana"] = 4
```

- **Adding elements:** Use a new key with square brackets and assign a value to it.

```
my_dict["grape"] = 5
```

- **Removing elements:** Use the pop() method to remove an element by key.

```
my_dict.pop("orange")
```

- **Iterating over elements:** Use the items() method to iterate over key-value pairs, the keys() method to iterate over keys, and the values() method to iterate over values.

```
for key, value in my_dict.items(): print(key, value)
```

Understanding and mastering these built-in data structures will allow you to be well-equipped to handle and manipulate data effectively in your Python programs. Practice using these data structures in different scenarios, and get comfortable with their methods and operations to enhance your problem-solving skills in Python.

Control Structures and Functions

Control structures are essential elements of programming that allow you to create more complex and dynamic code. They include conditional statements (if, else) and loops (for, while). In this section, we will discuss these control structures and introduce you to functions, which enable you to create reusable and modular code.

Conditional Statements (if-else)

Conditional statements allow you to execute different code blocks based on specific conditions. The most common conditional statement in Python is the if-else statement. The basic structure of an if-else statement is as follows:

```
if condition:
    # Code to execute if the condition is true else:
    # Code to execute if the condition is false
```

For example, let's create a simple program that checks whether a number is even or odd:

```
number = 7
if number % 2 == 0:
    print("The number is even.")
else:
    print("The number is odd.")
```

Loops (for, while)

Loops are used to execute a block of code repeatedly based on a condition. Python provides two types of loops: for and while.

A for loop is used to iterate over a sequence (such as a list, tuple, or string) and execute a block of code for each item in the sequence:

```
fruits = ["apple", "banana", "orange"]
for fruit in fruits:
    print(fruit)
```

A while loop, on the other hand, executes a block of code repeatedly as long as a given condition is true:

```
counter = 0
while counter < 5:
    print("Counter:", counter)
    counter += 1
```

Functions

Functions are blocks of reusable code that perform a specific task. They allow you to break your code into smaller, more manageable pieces, making it easier to understand and maintain. Functions are defined using the def keyword, followed by the function name and a pair of parentheses containing any input parameters. The function body is indented, and the return keyword is used to specify the output.

Here's an example of a simple function that calculates the square of a number:

```
def square(number):
    result = number ** 2
    return result
my_number = 5
my_square = square(my_number)
print("The square of", my_number, "is", my_square)
```

Understanding control structures and functions will allow you to create more complex and dynamic programs in Python. These skills are crucial when working with data, as they allow you to manipulate and process data efficiently. In the next section, we will explore Object-Oriented Programming in Python, which will further enhance your ability to write clean, reusable, and maintainable code in your Data Science projects.

String Manipulation

In this section, you will learn about various string manipulation techniques in Python, including string concatenation, slicing, formatting, and using built-in string methods. Mastering these techniques will enable you to work effectively with text data in your data science projects.

String Concatenation

Combine two or more strings into a single string using the + operator.

```
str1 = "Hello"
str2 = "World"
result = str1 + ", " + str2 print(result)
# Output: Hello, World
```

String Slicing

Extract a substring from a string by specifying the start and end indices. The start index is inclusive, while the end index is exclusive.

```
text = "Data Science"
substring = text[0:4] print(substring)
# Output: Data
```

You can also use negative indices to count from the end of the string, and omit the start or end index to slice from the beginning or to the end of the string.

```
last_four_chars = text[-4:] print(last_four_chars)
# Output: ence
```

String Formatting

Format strings by including placeholders within the string and providing the values to be inserted. One common method is to use f-strings, which are available in Python 3.6 and later versions. To create an f-string, add an f or F prefix to the string and use curly braces {} to enclose expressions.

```
name = "John"
age = 28
formatted_string = f"My name is {name} and I am {age} years old."
print(formatted_string)
# Output: My name is John and I am 28 years old.
```

Built-in String Methods

Python provides a variety of built-in string methods for common string manipulations. Here are some examples:

- **upper():** Convert all characters in a string to uppercase.

```
text = "hello world"
uppercase_text = text.upper() print(uppercase_text) # Output:
HELLO WORLD
```

- **lower():** Convert all characters in a string to lowercase.

```
text = "HELLO WORLD"
lowercase_text = text.lower() print(lowercase_text)
# Output: hello world
```

- **split():** Split a string into a list of substrings based on a specified delimiter. By default, the delimiter is any whitespace character.

```
text = "Data Science is amazing"
words = text.split()
```

```
print(words)
# Output: ['Data', 'Science', 'is', 'amazing']
```

- **join():** Join a list of strings into a single string using a specified delimiter.

```
words = ["Data", "Science", "is", "amazing"]
delimiter = " "
sentence = delimiter.join(words) print(sentence)
# Output: Data Science is amazing
```

Practice these string manipulation techniques and incorporate them into your Python programs to become proficient in handling and manipulating text data. These skills will be particularly useful when working with large volumes of textual data, such as during natural language processing tasks or when cleaning and processing datasets for your data science projects.

Error Handling

In this section, you will learn about the concept of error handling in Python using try-except blocks. Understanding error handling is essential for writing robust and reliable code that can handle unexpected situations gracefully.

Try-Except Blocks

Use try-except blocks to catch exceptions that may occur during the execution of your code. Enclose the code that might raise an exception within the try block, and handle the exception in the except block. If an exception is raised within the try block, the code in the except block will be executed.

```
try:
    numerator = 10
    denominator = 0
    result = numerator / denominator
except ZeroDivisionError:
    print("Error: Cannot divide by zero.")
```

Catching Specific Exceptions

Catch specific exceptions by mentioning the exception type in the except block. You can also use multiple except blocks to handle different exception types separately.

```
try:
    # code that might raise an exception
    pass
except ValueError:
    # handle ValueError
    pass
```

```
except TypeError:
    # handle TypeError
    pass
```

Creating Custom Exceptions

Create custom exception classes by inheriting from the built-in Exception class or one of its subclasses. This allows you to define your own exceptions that are more specific to your application or module.

```
class CustomError(Exception):
    def __init__(self, message):
        super().__init__(message)
try:
    # code that might raise an exception
    raise CustomError("This is a custom error.")
except CustomError as e:
    print(e) # Output: This is a custom error.
```

The finally Keyword

Use the finally keyword to include a block of code that will be executed no matter what, whether an exception is raised or not. This is useful for cleaning up resources, such as closing files or network connections, regardless of the outcome of the try-except block.

```
try:
    # code that might raise an exception
    pass
except ValueError:
    # handle ValueError
    pass
```

```
finally:
    print("This code will always be executed.")
```

Understanding and using error handling techniques in your Python code will allow you to create more reliable and robust programs that can handle unexpected situations gracefully. This is particularly important in data science projects, where dealing with real-world data often involves encountering edge cases and unexpected errors.

List Comprehensions

In this section, you will learn about list comprehensions, a concise and efficient way to create lists in Python. List comprehensions provide a shorter syntax for creating lists compared to traditional for loops, making your code more readable and often faster to execute.

Syntax

The basic syntax of a list comprehension is as follows:

```
[expression for item in iterable if condition]
```

This will create a new list by evaluating the expression for each item in the iterable that meets the condition. The if condition part is optional; if omitted, the list comprehension will iterate through all items in the iterable.

Examples

- Create a list of the squares of integers from 0 to 9:

```
squares = [x**2 for x in range(10)]
print(squares)
# Output: [0, 1, 4, 9, 16, 25, 36, 49, 64, 81]
```

- Create a list of even numbers from 1 to 20:

```
even_numbers = [num for num in range(1, 21) if num % 2 == 0]
print(even_numbers)
# Output: [2, 4, 6, 8, 10, 12, 14, 16, 18, 20]
```

- Convert a list of temperatures from Celsius to Fahrenheit:

```
celsius = [0, 10, 20, 30, 40]
fahrenheit = [(temp * 9/5) + 32 for temp in celsius]
print(fahrenheit)
# Output: [32.0, 50.0, 68.0, 86.0, 104.0
```

Comparison to For Loops

The following for loop can be used to create a list of the squares of integers from 0 to 9:

```
squares = []
for x in range(10):
    squares.append(x**2)
print(squares)
# Output: [0, 1, 4, 9, 16, 25, 36, 49, 64, 81]
```

This can be achieved more concisely using a list comprehension:

```
squares = [x**2 for x in range(10)]
print(squares)
# Output: [0, 1, 4, 9, 16, 25, 36, 49, 64, 81]
```

List comprehensions offer a more readable and efficient way to create lists in Python, especially when the logic is simple and can be expressed in a single line. By mastering list comprehensions, you will be able to write more concise and efficient code, which is particularly valuable in data science projects where manipulating and transforming large datasets is common.

File I/O

In this section, you will learn how to read and write files in Python, including opening and closing files, reading and writing text or binary data, and working with file paths. Being able to work with files is essential for many data science tasks, as data is often stored in files or needs to be saved to files for further processing or sharing.

Opening and Closing Files

Use the built-in open() function to open a file in Python. The function takes two arguments: the file path and the mode in which to open the file. The most common modes are:

- 'r': Read mode, for reading the contents of a file (default)
- 'w': Write mode, for creating a new file or overwriting an existing one
- 'a': Append mode, for appending data to an existing file
- 'b': Binary mode, for reading or writing binary data

Always close the file after you're done using it by calling the close() method on the file object. Alternatively, use the with statement to automatically close the file when the block of code is done executing.

```
# Opening a file in read mode
with open("example.txt", "r") as file:
# Perform file operations here pass
```

Reading and Writing Text Data

- Reading the entire file content:

```
with open("example.txt", "r") as file:
```

```
    content = file.read()
    print(content)
```

- Reading a single line at a time:

```
with open("example.txt", "r") as file:
    line = file.readline()
    print(line)
```

- Reading all lines into a list:

```
with open("example.txt", "r") as file:
    lines = file.readlines()
    print(lines)
```

- Writing data to a file:

```
with open("output.txt", "w") as file:
    file.write("Hello, World!")
```

Reading and Writing Binary Data

- Reading binary data:

```
with open("example.bin", "rb") as file:
    binary_data = file.read()
    print(binary_data)
```

- Writing binary data:

```
binary_data =
b'\x48\x65\x6c\x6c\x6f\x2c\x20\x57\x6f\x72\x6c\x64\x21' with
open("output.bin", "wb") as file:
    file.write(binary_data)
```

Working with File Paths

Use the os module to work with file paths in a platform-independent manner. This ensures that your code works correctly on different operating systems, such as Windows, macOS, and Linux.

```
import os


# Join file path components
file_path = os.path.join("my_folder", "example.txt")
print(file_path)
# Output: "my_folder/example.txt" on Unix-based systems


# Check if a file exists
file_exists = os.path.exists(file_path)
print(file_exists)
# Output: True or False


# Get the directory name and file name from a file path
dir_name = os.path.dirname(file_path)
file_name = os.path.basename(file_path)
print(dir_name) # Output: "my_folder"
print(file_name) # Output: "example.txt"
```

Understanding and using these file I/O techniques will allow you to be well-equipped to handle data stored in files and save the results of your data processing tasks. This is a crucial skill for data scientists, as working with files is often an integral part of data acquisition, preprocessing, and sharing results with others.

Modules and Packages

In this section, you will learn about modules and packages in Python. Understanding modules and packages is crucial for organizing and reusing your code efficiently, as well as for leveraging the vast ecosystem of third-party libraries available for Python.

Modules

A module is a single Python file containing code, such as functions, classes, and variables, which can be imported into other Python scripts to reuse functionality. To create a module, simply save your code in a file with a '.py' extension.

- Creating a module:

```
# my_module.py
def greet(name):
    print(f"Hello, {name}!")
pi = 3.14159
```

- Importing a module:

To use the functionality of a module in another Python script, use the import statement, followed by the module name without the '.py' extension.

```
# main.py
import my_module

my_module.greet("John")
print(my_module.pi)
```

Packages

A package is a collection of related modules organized in a directory hierarchy. To create a package, create a directory with an __init__.py file, which can be empty or contain package-level code. Place your module files in this directory.

- Creating a package:

```
my_package/
    __init__.py
    module1.py
    module2.py
```

- Importing a package or a specific module from a package:

To import a package or a specific module from a package, use the import statement followed by the package name or the package and module name separated by a dot.

```
# Import the entire package
    import my_package

# Import a specific module from the package
import my_package.module1

# Import a specific function or variable from a module in the
package
from my_package.module1 import my_function, my_variable
```

Aliases and selective imports

You can create aliases for imported modules, functions, or variables using the as keyword. This can help make your code more readable by using shorter or more descriptive names.

```
# Import a module with an alias
```

```
import my_package.module1 as m1

# Import a function with an alias
from my_package.module1 import my_function as func
```

Understanding and using modules and packages will allow you to better organize your code, making it more maintainable and reusable. Additionally, you will be able to leverage the vast ecosystem of Python libraries and packages, such as NumPy, Pandas, and Matplotlib, to enhance your data science projects with powerful, pre-built functionality.

Creating your own packages

Creating your own package is an excellent way to organize and reuse code across multiple projects. Here's a step-by-step guide on how to create your own Python package:

1. Choose a descriptive name for your package and create a directory with that name. This will be the top-level directory for your package. Make sure the name does not conflict with existing Python packages or modules.

```
my_package/
```

2. Inside the top-level directory, create an __init__.py file. This file is required for Python to treat the directory as a package. It can be empty or contain package-level code, such as constants or functions that are common to all modules in the package.

```
my_package/
    __init__.py
```

3. Add module files (.py files) to the package directory. These files should contain the code for your package, such as functions, classes, and variables. Organize your code into separate modules based on their functionality.

```
my_package/
    __init__.py
    module1.py
    module2.py
```

4. (Optional) Include a README.md file in the package directory to provide documentation and instructions on how to use your package. This is particularly helpful if you plan to share your package with others or publish it on a platform like GitHub.

5. (Optional) Add any necessary sub-packages or subdirectories, following the same structure as described in steps 2 and 3. Sub-packages can help you further organize your code if your package becomes large and complex.

```
my_package/
    __init__.py
    module1.py
    module2.py
    sub_package/
        __init__.py
        sub_module1.py
        sub_module2.py
```

6. To use your package in another Python script or project, add the package directory to your PYTHONPATH environment variable or place the package directory in a location that Python can find, such as the site-packages

directory of your Python installation. Then, you can import your package or its modules just like any other Python package:

```
import my_package
import my_package.module1
from my_package.module2 import my_function
```

Creating your own package will allow you to effectively organize, reuse, and share your code with others. This skill is particularly valuable in data science, where modular, well-organized code can significantly improve the maintainability and extensibility of your projects.

Working with Dates and Times

In this section, you will learn about Python's datetime module, which provides classes and functions for working with dates, times, and time intervals. Understanding how to create, format, and manipulate dates and times is essential in data science, as many datasets include time-related information that needs to be processed, analyzed, or presented.

Creating Dates and Times

The datetime module provides several classes for representing dates, times, and their combinations:

- date: Represents a date (year, month, and day) without time information.
- time: Represents a time of day (hour, minute, second, and microsecond) without date information.
- datetime: Represents a date and time.

To create instances of these classes, use their respective constructors:

```
from datetime import date, time, datetime


# Create a date
my_date = date(2023, 5, 8)
print(my_date) # Output: 2023-05-08


# Create a time
my_time = time(12, 30, 45)
print(my_time)
# Output: 12:30:45
```

```
# Create a datetime
my_datetime = datetime(2023, 5, 8, 12, 30, 45)
print(my_datetime)
# Output: 2023-05-08 12:30:45
```

Current Date and Time

Use the following methods to obtain the current date and time:

- date.today(): Returns the current date.
- datetime.now(): Returns the current date and time.

```
current_date = date.today()
print(current_date)

current_datetime = datetime.now()
print(current_datetime)
```

Formatting Dates and Times

Use the strftime() method to format dates, times, or datetimes as strings, with various format codes representing different components:

- %Y: Year with century (e.g., 2023)
- %m: Month as a zero-padded decimal number (e.g., 05)
- %d: Day of the month as a zero-padded decimal number (e.g., 08)
- %H: Hour (24-hour clock) as a zero-padded decimal number (e.g., 12)
- %M: Minute as a zero-padded decimal number (e.g., 30)
- %S: Second as a zero-padded decimal number (e.g., 45)

```
formatted_date = my_datetime.strftime("%Y-%m-%d")
print(formatted_date)
# Output: 2023-05-08

formatted_time = my_datetime.strftime("%H:%M:%S")
print(formatted_time)
# Output: 12:30:45

formatted_datetime = my_datetime.strftime("%Y-%m-%d %H:%M:
%S")
print(formatted_datetime)
# Output: 2023-05-08 12:30:45
```

Parsing Dates and Times

Use the strptime() method to parse a date, time, or datetime from a string, providing a format string that matches the input:

```
input_str = "2023-05-08 12:30:45"
parsed_datetime = datetime.strptime(input_str, "%Y-%m-%d %H:
%M:%S")
print(parsed_datetime)
# Output: 2023-05-08 12:30:45
```

Manipulating Dates and Times

Use the timedelta class to represent a duration or time interval, which can be added to or subtracted from dates or datetimes:

```
from datetime import timedelta

# Create a timedelta representing 10 days, 2 hours, and 30
minutes
duration = timedelta(days=10, hours=2, minutes=30)
```

```
# Add the timedelta to a date or datetime
new_date = my_date + duration
print(new_date) # Output: 2023-05-18
new_datetime = my_datetime + duration
print(new_datetime)
# Output: 2023-05-18 15:01:45

# Subtract the timedelta from a date or datetime
previous_date = new_date - duration
print(previous_date) # Output: 2023-05-08
previous_datetime = new_datetime - duration
print(previous_datetime)
# Output: 2023-05-08 12:30:45
```

Comparing Dates and Times

Dates, times, and datetimes can be compared using standard comparison operators (`<`, `>`, `==`, `!=`, `<=`, `>=`):

```
date1 = date(2023, 5, 8)
date2 = date(2023, 6, 1)

# Check if date1 is earlier than date2
print(date1 < date2)
# Output: True

# Check if two datetimes are equal
datetime1 = datetime(2023, 5, 8, 12, 30, 45)
datetime2 = datetime(2023, 5, 8, 12, 30, 45)
print(datetime1 == datetime2)
```

```
# Output: True
```

Mastering the datetime module and its various classes and functions will allow you to be well-equipped to handle a wide range of time-related tasks in your data science projects. This knowledge will enable you to process, analyze, and present time-based data efficiently and accurately.

Lambda Functions

In this section, you will learn about lambda functions, which are small, anonymous functions defined using the lambda keyword. Lambda functions can be useful in situations where you need a simple function for a short period of time and don't want to define a full function using the def keyword. We will discuss their use cases, syntax, and limitations compared to regular functions.

Syntax

Lambda functions have a concise syntax that consists of the lambda keyword, followed by a comma-separated list of arguments, a colon, and an expression. The lambda function can accept any number of arguments but must have only one expression.

```
lambda arguments: expression
```

Use Cases

Lambda functions are commonly used for simple operations or as arguments for higher-order functions, such as map(), filter(), or sorted(), which expect a function as one of their inputs.

Examples

- Sorting a list of tuples based on the second element:

```
data = [(1, 'B'), (2, 'A'), (3, 'C')]
sorted_data = sorted(data, key=lambda x: x[1])
print(sorted_data)
# Output: [(2, 'A'), (1, 'B'), (3, 'C')]
```

- Squaring each element in a list using map():

```
numbers = [1, 2, 3, 4, 5]
```

```
squares = list(map(lambda x: x**2, numbers))
print(squares)
# Output: [1, 4, 9, 16, 25]
```

- Filtering out even numbers from a list using filter():

```
numbers = [1, 2, 3, 4, 5]
odd_numbers = list(filter(lambda x: x % 2 != 0, numbers))
print(odd_numbers)
# Output: [1, 3, 5]
```

Limitations

While lambda functions are concise and convenient for simple operations, they have some limitations compared to regular functions:

- Lambda functions can only have a single expression, making them unsuitable for more complex operations or multi-line functions.
- They are anonymous, which means they don't have a name and cannot be easily reused or referred to in other parts of your code.
- They can't include statements, assignments, or multiple expressions, which can make them less readable for more complex operations.

In cases where a more complex operation is needed or a function will be reused multiple times, it's better to use a regular function defined with the def keyword.

Understanding when and how to use lambda functions will allow you to write more concise and efficient code in certain situations. However, it's important to recognize their limitations and know when to use a regular function instead. Balancing the use of lambda functions and regular functions will help you write clear, maintainable, and efficient code in your data science projects.

Object-Oriented Programming in Python

In this section, you will learn about Object-Oriented Programming (OOP) in Python, which is a programming paradigm that uses objects and classes to model real-world entities and their interactions. OOP allows you to write modular, reusable, and maintainable code, which is essential for data science projects.

Classes and Objects

A class is a blueprint for creating objects (instances) with specific attributes and methods. An object is an instance of a class, and it represents a specific entity with its own set of attributes and behaviors defined by the class.

To define a class in Python, use the class keyword, followed by the class name:

```
class MyClass:
    pass
```

To create an object from a class, call the class like a function:

```
my_object = MyClass()
```

Attributes and Methods

Attributes are variables associated with a class or its objects, while methods are functions associated with a class or its objects. Use the __init__ method to initialize an object's attributes when it's created:

```
class Car:
    def __init__(self, make, model, year):
```

```
        self.make = make
        self.model = model
        self.year = year
```

To define a method, create a function within the class, with the first parameter being self:

```
class Car:
    # ...
    def get_description(self):
        return f"{self.year} {self.make} {self.model}"
```

To access an object's attributes or methods, use the dot notation:

```
my_car = Car("Tesla", "Model 3", 2021)
print(my_car.make) # Output: Tesla
print(my_car.get_description())
# Output: 2021 Tesla Model 3
```

Inheritance

Inheritance allows you to create a new class that inherits attributes and methods from an existing class. The new class is called a subclass or derived class, and the existing class is called the superclass or base class.

To create a subclass, specify the superclass in parentheses after the class name:

```
class ElectricCar(Car):
    def __init__(self, make, model, year, battery_size):
    super().__init__(make, model, year)
    self.battery_size = battery_size
```

The super() function allows you to call a method from the superclass, such as __init__ in this case.

Polymorphism

Polymorphism is the ability to use a single interface to represent different types. In Python, this can be achieved through method overriding, where a subclass provides its own implementation of a method that already exists in the superclass.

```
class ElectricCar(Car): # ... def get_description(self): return
f"{self.year} {self.make} {self.model} - {self.battery_size} kWh
battery"
```

Understanding the principles of Object-Oriented Programming in Python will allow you to write more modular, reusable, and maintainable code in your data science projects. This will help you create more efficient and organized code, making it easier to debug, modify, and scale your projects as needed.

Module 2 Exam Questions

1. What does cybersecurity refer to?

a) The practice of protecting computers, networks, applications, and data from unauthorized access, theft, or damage

b) The process of creating secure passwords

c) The use of firewalls and antivirus software

d) The study of cybercriminals and their motives

1. What is the main purpose of using lambda functions in Python?

a) For complex multi-line functions

b) For simple one-line functions

c) To define a new class

d) To handle file I/O operations

2. How do you define a class in Python?

a) def ClassName():

b) class ClassName:

c) function ClassName:

d) class ClassName()

3. What method is used to initialize an object's attributes in Python?

a) init()

b) start()

c) begin()

d) create()

4. Which Python data structure is used to store multiple values in a single variable, and allows duplicate values?

a) Set

b) List

c) Tuple

d) Dictionary

5. What is the output of this code: 'Data Science'.split()

a) ['Data', 'Science']

b) 'Data Science'

c) 'DataScience'

d) None of the above

6. What method is used to get the length of a list in Python?

a) length()

b) count()

c) size()

d) len()

7. Which of the following Python libraries is used for scientific computing and working with arrays?

a) Pandas

b) NumPy

c) Matplotlib

d) Seaborn

8. What is the purpose of the map() function in Python?

a) To apply a function to all elements in a list

b) To create a new list with specific elements from an existing list

c) To filter elements from a list based on a condition

d) To sort a list based on a custom key function

9. What do you use to read and write files in Python?

a) File I/O

b) Lambda functions

c) List comprehensions

d) Error handling

10. Which of the following is not a valid Python data type?

a) List

b) Array

c) Tuple

d) Set

11. In Python, what does the term "list comprehension" refer to?

a) A concise way to create lists

b) A way to handle errors in list operations

c) A way to manipulate strings

d) A way to define a new class

12. Which of the following is a classification algorithm in machine learning?

a) Linear regression

b) K-means clustering

c) Decision trees

d) Principal component analysis

13. What is the primary goal of data visualization in data science?

a) To present data in a clear and informative manner

b) To preprocess data for machine learning

c) To perform statistical analysis

d) To create Python classes and objects

14. In Python, what is the result of this operation: [1, 2] + [3, 4]?

a) [1, 2, 3, 4]

b) [4, 6]

c) [1, 2, 6]

d) An error

15. What is the primary use of the try and except blocks in Python?

a) To create classes and objects

b) To handle errors and exceptions

c) To create lambda functions

d) To manipulate strings

16. What is the output of this code: 'Hello, {}'.format('World')

a) 'Hello, World'

b) 'Hello, {World}'

c) 'Hello, {}World'

d) None of the above

17. Which of the following is a Python built-in function for working with dates and times?

a) date

b) datetime

c) time

d) All of the above

Answers:

1. b
2. b
3. a
4. b
5. a
6. d
7. b
8. a
9. a
10. b
11. a
12. c
13. a
14. a
15. b
16. a
17. b
18. c
19. b
20. c

Module 3. Python Libraries for Data Science

Introduction to Python Libraries for Data Science

Welcome to Module 3 on Python Libraries for Data Science. In this module, you will learn about some of the most powerful and widely used libraries in Python for data analysis, manipulation, and visualization. These libraries have been developed by the data science community over the years to simplify and streamline various tasks, allowing you to focus on solving problems and drawing insights from your data.

Python has become a popular choice among data scientists, not only because of its ease of use and readability but also due to the vast ecosystem of libraries available. These libraries enable you to perform a wide range of tasks, from basic data manipulation to advanced machine learning techniques, with minimal coding.

In this module, we will introduce you to three essential Python libraries: NumPy, Pandas, and Matplotlib with Seaborn. You will learn the fundamental concepts behind each library, their unique features, and how to harness their capabilities to process and analyze your data efficiently.

For instance, NumPy, a library focused on numerical computing, provides a robust array structure and an assortment of mathematical functions that allow you to perform complex calculations with ease. A practical example of using NumPy could be calculating the mean, median, and standard deviation of a dataset, which are essential metrics for understanding the distribution of your data.

Pandas is another powerful library that builds on top of NumPy and offers a more sophisticated data structure called DataFrame, which is specifically designed for data manipulation and analysis. With Pandas, you can effortlessly handle large datasets, perform data cleaning, and merge or join different tables. A practical example of using Pandas would be analyzing a real-world dataset containing sales records, filtering out invalid entries, and aggregating the results to understand overall trends.

Lastly, we will explore Matplotlib and Seaborn, two libraries that enable you to create stunning and informative visualizations to better understand your data and share your findings with others. Through practical examples, you will learn how to create different types of plots, such as bar charts, line plots, and scatterplots, to uncover patterns and relationships within your data.

By the end of this module, you will have a solid understanding of these key Python libraries and their applications in data science. You will be equipped with the skills to confidently utilize these libraries in your data analysis projects, enabling you to transform raw data into meaningful insights.

NumPy

NumPy, short for Numerical Python, is a foundational library for scientific computing in Python. It provides a powerful and efficient N-dimensional array object, which allows you to perform mathematical operations on large datasets with ease. NumPy also offers a wide range of mathematical functions that are optimized for performance, making it an essential tool for your data science toolkit.

In this section, we will introduce you to the core features of NumPy and show you how to effectively utilize its capabilities for your data analysis tasks.

Creating Arrays

NumPy arrays are the backbone of the library and are used to store and manipulate data. You can create NumPy arrays from lists, tuples, or other array-like objects using the numpy.array() function. Here's an example of creating a simple NumPy array:

```
import numpy as np my_array = np.array([1, 2, 3, 4, 5])
print(my_array)
```

Array Attributes

NumPy arrays have several useful attributes that provide information about the array, such as its shape, size, and data type. For example:

```
print(my_array.shape) # Output: (5,)
print(my_array.size) # Output: 5
print(my_array.dtype) # Output: int64
```

Multidimensional Arrays

NumPy allows you to work with multidimensional arrays, which are useful for representing complex data structures like matrices. You can create a multidimensional array by passing nested lists or tuples to the numpy.array() function:

```
my_2d_array = np.array([[1, 2, 3], [4, 5, 6], [7, 8, 9]])
print(my_2d_array)
```

Array Operations

You can perform various mathematical operations on NumPy arrays, such as element-wise addition, subtraction, multiplication, and division. For example:

```
array1 = np.array([1, 2, 3])
array2 = np.array([4, 5, 6])
sum_array = array1 + array2
print(sum_array) # Output: [5, 7, 9]
```

Indexing and Slicing

Similar to lists in Python, you can access individual elements of a NumPy array using indexing and slicing techniques. For example:

```
my_array = np.array([1, 2, 3, 4, 5])
# Access the first element
print(my_array[0]) # Output: 1
# Access elements from index 1 to 3 (excluding 3)
print(my_array[1:3]) # Output: [2, 3]
```

Mathematical Functions

NumPy provides an extensive collection of mathematical functions that you can use to perform calculations on your arrays. Some common functions include:

- numpy.sum(): Calculate the sum of elements in an array.
- numpy.mean(): Compute the mean (average) of an array.
- numpy.median(): Find the median of an array.
- numpy.std(): Calculate the standard deviation of an array.

Here's an example of using these functions on an array:

```
my_array = np.array([1, 2, 3, 4, 5])
print(np.sum(my_array)) # Output: 15
print(np.mean(my_array)) # Output: 3.0
print(np.median(my_array)) # Output: 3.0
print(np.std(my_array)) # Output: 1.41421356
```

Array Broadcasting

One of the powerful features of NumPy is array broadcasting, which allows you to perform operations on arrays of different shapes and sizes. Broadcasting automatically adjusts the dimensions of the arrays involved in the operation to make them compatible. This feature simplifies your code and improves performance by minimizing the need for explicit loops.

For example, suppose you want to add a constant value to each element of an array:

```
my_array = np.array([1, 2, 3, 4, 5])
constant = 10
result = my_array + constant
print(result) # Output: [11, 12, 13, 14, 15]
```

In this case, the constant value is broadcasted to match the shape of my_array, allowing you to perform the addition operation seamlessly.

Array Reshaping

Reshaping arrays is a common operation in data manipulation and preprocessing. NumPy provides the reshape() function, which allows you to change the shape of an array without altering its data. For example, you can convert a 1-dimensional array into a 2-dimensional array (matrix) or vice versa:

```
my_array = np.array([1, 2, 3, 4, 5, 6])
# Reshape the array into a 2x3 matrix
my_matrix = my_array.reshape(2, 3)
print(my_matrix) # Output: # [[1, 2, 3], # [4, 5, 6]]
```

Array Concatenation and Splitting

NumPy allows you to concatenate multiple arrays along a specified axis using the numpy.concatenate() function. This is useful when you need to combine data from different sources or merge the results of multiple operations. For example:

```
array1 = np.array([1, 2, 3])
array2 = np.array([4, 5, 6])
# Concatenate arrays along the first axis
combined_array = np.concatenate((array1, array2))
print(combined_array) # Output: [1, 2, 3, 4, 5, 6]
```

You can also split an array into multiple smaller arrays using the numpy.split() function:

```
my_array = np.array([1, 2, 3, 4, 5, 6])
# Split the array into 3 equal parts
split_arrays = np.split(my_array, 3)
print(split_arrays) # Output: [array([1, 2]), array([3, 4]), array([5, 6])]
```

Linear Algebra Operations

NumPy offers a comprehensive suite of linear algebra functions, which can be helpful for solving systems of equations, finding eigenvalues and eigenvectors, and performing matrix decompositions. Some of the key linear algebra functions include:

- numpy.dot(): Compute the dot product of two arrays.
- numpy.linalg.inv(): Calculate the inverse of a square matrix.
- numpy.linalg.eig(): Find the eigenvalues and eigenvectors of a square matrix.
- numpy.linalg.svd(): Perform singular value decomposition on a matrix.

For example, to find the dot product of two matrices, you can use the following code:

```
matrix1 = np.array([[1, 2], [3, 4]])
matrix2 = np.array([[5, 6], [7, 8]])
dot_product = np.dot(matrix1, matrix2)
print(dot_product) #Output: [[19, 22], [43, 50]]
```

Random Number Generation

NumPy provides a suite of functions for generating random numbers, which are useful for simulating data or initializing the weights in machine learning models. The `numpy.random` module contains functions for generating random numbers from various distributions, such as uniform, normal, and binomial distributions. Some common random number generation functions include:

- numpy.random.rand(): Generate random numbers from a uniform distribution between 0 and 1.
- numpy.random.randn(): Generate random numbers from a standard normal distribution (mean 0, standard deviation 1).

- numpy.random.randint(): Generate random integers within a specified range.

For example, to generate a 3x3 matrix of random numbers from a standard normal distribution, you can use the following code: random_matrix = np.random.randn(3, 3) print(random_matrix)

Advanced Indexing and Masking

NumPy allows you to use advanced indexing techniques to select specific elements or subsets of an array based on conditions. For example, you can use boolean indexing to filter an array based on a condition:

```
my_array = np.array([1, 2, 3, 4, 5, 6])
# Create a boolean mask for elements greater than 3
mask = my_array > 3 print(mask)
# Output: [False, False, False, True, True, True]

# Select elements greater than 3 using the boolean mask
filtered_array = my_array[mask]
print(filtered_array) # Output: [4, 5, 6]
```

You can also use integer arrays to index another array, which is useful for rearranging or selecting elements in a specific order:

```
my_array = np.array([10, 20, 30, 40, 50])
# Index array with the desired order of elements
index_array = np.array([4, 0, 1, 3, 2])

# Rearrange the elements using integer indexing
rearranged_array = my_array[index_array]
print(rearranged_array) # Output: [50, 10, 20, 40, 30]
```

In summary, the NumPy library offers a vast array of features and functions that make numerical computing in Python efficient and versatile. As you delve deeper into data science, you'll continue to find NumPy invaluable for tasks such as data preprocessing, statistical analysis, and machine learning. By mastering NumPy's capabilities, you'll be well-equipped to tackle complex data analysis challenges and create robust, efficient solutions.

Pandas

Pandas is a powerful and versatile Python library designed for data manipulation and analysis. It provides two main data structures: the DataFrame and the Series, which are built on top of NumPy arrays. These structures make it easy to handle large datasets, perform data cleaning, and merge or join different tables. In this section, we will introduce you to the core features of Pandas and show you how to effectively utilize its capabilities for your data analysis tasks.

Creating DataFrames

A DataFrame is a two-dimensional table that can store data of various types, such as integers, floats, and strings. You can create DataFrames from dictionaries, lists, NumPy arrays, or even by reading data from external files, such as CSV or Excel files. Here's an example of creating a simple DataFrame from a dictionary:

```
import pandas as pd
data = {
    'Name': ['Alice', 'Bob', 'Carol', 'David'],
    'Age': [25, 30, 35, 40],
    'City': ['New York', 'San Francisco', 'Los Angeles', 'Chicago']
}
df = pd.DataFrame(data)
print(df)
```

Accessing Data in DataFrames

You can access columns in a DataFrame using the column name, like this:

```
age_column = df['Age']
print(age_column)
```

To access specific rows or subsets of the DataFrame, you can use the iloc[] and loc[] attributes. The iloc[] attribute allows you to access rows by their integer index, while loc[] enables you to access rows by their index labels:

```
# Access the second row using iloc[]
row_1 = df.iloc[1]
print(row_1)

# Access rows with index labels 1 and 2 using loc[]
rows_1_2 = df.loc[1:2]
print(rows_1_2)
```

Manipulating Data in DataFrames

Pandas provides a wide range of functions for modifying and transforming your data. Some common operations include adding new columns, updating values, and filtering rows based on conditions. For example, to add a new column to a DataFrame, you can simply assign a new Series or list to a new column name:

```
df['Country'] = ['USA', 'USA', 'USA', 'USA']
print(df)
```

To filter a DataFrame based on a condition, you can use boolean indexing, similar to NumPy:

```
# Filter rows where Age is greater than 30
filtered_df = df[df['Age'] > 30]
print(filtered_df)
```

Handling Missing Data

Pandas provides robust tools for handling missing data, which is a common challenge in data analysis. You can use the isnull() and notnull() functions to detect missing values in your DataFrame. To fill or drop missing values, you can use the fillna() and dropna() methods, respectively:

```
# Fill missing values with a specified value
filled_df = df.fillna(value=0) # Drop rows with missing values
clean_df = df.dropna()
```

Grouping and Aggregating Data

Grouping and aggregating data is an essential part of data analysis, as it allows you to uncover patterns and relationships within your data. With Pandas, you can use the groupby() function to group your data based on one or more columns, and then apply aggregation functions such as sum(), mean(), or count() to compute summary statistics for each group:

```
# Group data by 'City' and compute the mean 'Age'
mean_age_by_city = df.groupby('City')['Age'].
mean()
print(mean_age_by_city)
```

Merging and Joining DataFrames

In many data analysis tasks, you'll need to combine data from multiple sources or tables. Pandas provides several methods for merging and joining DataFrames, such as merge(), join(), and concat():

- merge(): Merge two DataFrames based on a common column or index.
- join(): Join two DataFrames based on their index values.

- concat(): Concatenate multiple DataFrames along a specified axis.

For example, to merge two DataFrames based on a common column, you can use the following code:

```
df1 = pd.DataFrame({
    'Name': ['Alice', 'Bob', 'Carol', 'David'],
    'Age': [25, 30, 35, 40]
})
df2 = pd.DataFrame({
    'Name': ['Alice', 'Bob', 'Carol', 'David'],
    'City': ['New York', 'San Francisco', 'Los Angeles', 'Chicago']
})
# Merge df1 and df2 on the 'Name' column
merged_df = pd.merge(df1, df2, on='Name')
print(merged_df)
```

Reading and Writing Data from External Files

Pandas makes it easy to read and write data from various file formats, such as CSV, Excel, JSON, and SQL databases. Some commonly used functions for reading and writing data include:

- pd.read_csv(): Read data from a CSV file.
- pd.read_excel(): Read data from an Excel file.
- pd.read_json(): Read data from a JSON file.
- DataFrame.to_csv(): Write DataFrame data to a CSV file.
- DataFrame.to_excel(): Write DataFrame data to an Excel file.
- DataFrame.to_json(): Write DataFrame data to a JSON file.

For example, to read data from a CSV file and create a DataFrame, you can use the following code:

```
df = pd.read_csv('my_data.csv') print(df)
```

To write your DataFrame data to a new CSV file, you can use the to_csv() method:

```
df.to_csv('my_new_data.csv', index=False)
```

Time Series Analysis

Pandas provides powerful tools for working with time series data, such as date ranges, resampling, and time zone handling. You can easily create date ranges using the pd.date_range() function and use these as index values in your DataFrames:

```
# Create a date range of daily timestamps
date_range = pd.date_range(start='2020-01-01',
end='2020-12-31', freq='D')
# Create a DataFrame with the date range as the index
time_series_df = pd.DataFrame(index=date_range)
```

Resampling is a common operation in time series analysis, where you change the frequency of your data, such as converting daily data to monthly data. You can use the resample() method to achieve this:

```
# Resample daily data to monthly data and compute the mean
value for each month
monthly_data = time_series_df.resample('M').mean()
```

In summary, Pandas is a versatile and powerful library that enables you to efficiently manipulate, clean, and analyze data in Python. By mastering the core features and functions of Pandas, you'll be able to tackle complex data analysis tasks with ease and confidence. As you continue to explore the world of data science, you'll find Pandas to be an indispensable tool for handling data and uncovering insights.

Matplotlib and Seaborn

Visualizing data is a critical aspect of data analysis, as it allows you to explore patterns, relationships, and trends within your data. In this section, we will introduce you to two powerful Python libraries for data visualization: Matplotlib and Seaborn. These libraries offer a wide range of tools and techniques for creating informative and engaging visualizations, which will help you effectively communicate your insights and findings.

Matplotlib

Matplotlib is a widely-used data visualization library in Python that provides a flexible and versatile interface for creating static, animated, and interactive visualizations. It supports various types of plots, such as line plots, scatter plots, bar plots, and histograms, to name just a few. To get started with Matplotlib, you'll first need to import the library:

```
import matplotlib.pyplot as plt
```

Here are some examples of creating basic plots with Matplotlib:

- **Line Plot:**

```
import numpy as np
# Create sample data
x = np.linspace(0, 10, 100)
y = np.sin(x)
# Create a line plot of y vs. x
plt.plot(x, y)
# Add labels and a title
plt.xlabel('x-axis')
plt.ylabel('y-axis')
```

```
plt.title('Sine Wave')
# Display the plot
plt.show()
```

- **Scatter Plot:**

```
# Create sample data
x = np.random.rand(50)
y = np.random.rand(50)
# Create a scatter plot of y vs. x
plt.scatter(x, y)
# Add labels and a title
plt.xlabel('x-axis')
plt.ylabel('y-axis')
plt.title('Scatter Plot')
# Display the plot
plt.show()
```

- **Bar Plot:**

```
# Create sample data
categories = ['A', 'B', 'C', 'D']
values = [10, 20, 30, 40]
# Create a bar plot
plt.bar(categories, values)
# Add labels and a title
plt.xlabel('Category')
plt.ylabel('Value')
plt.title('Bar Plot')
# Display the plot
plt.show()
```

Seaborn

Seaborn is a statistical data visualization library built on top of Matplotlib that provides a high-level interface for creating elegant and informative visualizations. It comes with several built-in themes and color palettes that make it easy to create aesthetically pleasing plots. Seaborn also offers a variety of advanced plotting functions, such as violin plots, box plots, and pair plots, which can help you uncover deeper insights into your data. To get started with Seaborn, you'll need to import the library:

```
import seaborn as sns
```

Here are some examples of creating plots with Seaborn:

- **Violin Plot:**

```
import pandas as pd
# Create sample data
data = {
    'Category': ['A', 'A', 'B', 'B', 'C', 'C', 'D', 'D'],
    'Value': [10, 15, 20, 25, 30, 35, 40, 45]
}
df = pd.DataFrame(data)
# Create a violin plot
sns.violinplot(x='Category', y='Value', data=df)
# Display the plot
plt.show()
```

- **Box Plot:**

```
# Create a box plot
sns.boxplot(x='Category', y='Value', data=df)
# Display the plot
```

```
plt.show()
```

- **Pair Plot:**

```
# Create a pair plot of a DataFrame
sns.pairplot(df, hue='Category')
# Display the plot
plt.show()
```

In summary, Matplotlib and Seaborn are essential libraries for data visualization in Python. They provide a wide range of tools and functions for creating both simple and complex visualizations that can help you explore, analyze, and present your data effectively.

Mastering the core features of Matplotlib and Seaborn will allow you to create informative and engaging visualizations that can aid in your data analysis tasks and improve your ability to communicate your findings. As you continue to work with data in Python, you'll find that these libraries are invaluable tools for gaining deeper insights and understanding from your data. So, invest time in learning and experimenting with various plotting functions, styles, and techniques to elevate your data visualization skills to the next level.

Module 3 Exam Questions

1. What is the primary data structure used in Pandas for two-dimensional tabular data?

a) NumPy Array

b) List

c) DataFrame

d) Dictionary

2. Which Python library is primarily used for data manipulation and analysis?

a) Matplotlib

b) Pandas

c) Seaborn

d) NumPy

3. Which function is used to create a new NumPy array with specific dimensions?

a) np.array()

b) np.arange()

c) np.zeros()

d) np.ones()

4. In a Pandas DataFrame, how do you access the data in a specific column by its name?

a) df['column_name']

b) df.loc['column_name']

c) df.iloc['column_name']

d) df.at['column_name']

5. What type of plot is commonly used to visualize the distribution of a single variable?

a) Line plot

b) Scatter plot

c) Bar plot

d) Histogram

6. What is the purpose of the "iloc[]" attribute in a Pandas DataFrame?

a) Access rows by their index labels

b) Access columns by their column names

c) Access rows by their integer index

d) Access rows based on a condition

7. Which Python library is specifically designed for creating static, animated, and interactive visualizations?

a) NumPy

b) Pandas

c) Seaborn

d) Matplotlib

8. In NumPy, which function is used to perform element-wise addition of two arrays?

a) np.add()

b) np.subtract()

c) np.multiply()

d) np.divide()

9. What is the primary data structure used in NumPy for storing multi-dimensional data?

a) List

b) DataFrame

c) Dictionary

d) NumPy Array

10. Which Pandas function is used to read data from a CSV file and create a DataFrame?

a) pd.read_csv()

b) pd.read_excel()

c) pd.read_json()

d) pd.read_sql()

11. In Matplotlib, which function is used to display a created plot?

a) plt.show()

b) plt.plot()

c) plt.scatter()

d) plt.bar()

12. Which Pandas method is used to drop rows with missing values in a DataFrame?

a) fillna()

b) dropna()

c) isnull()

d) notnull()

13. How can you create a Seaborn box plot?

a) sns.violinplot()

b) sns.boxplot()

c) sns.scatterplot()

d) sns.lineplot()

14. Which function is used to create a NumPy array with evenly spaced values between a start and end value?

a) np.linspace()

b) np.arange()

c) np.zeros()

d) np.ones()

15. In a Pandas DataFrame, which method is used to compute summary statistics for each group after grouping data based on one or more columns?

a) aggregate()

b) summarize()

c) groupby()

d) pivot_table()

16. What is the primary purpose of the Seaborn library?

a) Data manipulation

b) Statistical data visualization

c) Advanced mathematical calculations

d) Machine learning algorithms

17. Which function is used to create a new Pandas DataFrame from a dictionary?

a) pd.DataFrame()

b) pd.Series()

c) pd.read_dict()

d) pd.create_frame()

18. Which method is used to resample time series data in Pandas?

a) groupby()

b) aggregate()

c) resample()

d) pivot_table()

19.How do you access a specific element in a NumPy array using its index?

a) array[index]

b) array.loc[index]

c) array.iloc[index]

d) array.at[index]

20. Which attribute in a Pandas DataFrame is used to access the index of the DataFrame?

a) df['index']

b) df.index

c) df.iloc[]

d) df.loc[]

Answers:

1. c
2. b
3. d
4. a
5. d
6. c
7. d
8. a
9. d
10. a
11. a
12. b
13. b
14. a
15. c
16. b
17. a
18. c
19. a
20. b

Module 4. Foundations of Statistics for Data Science

Introduction to Statistics for Data Science

Welcome to Module 4 on Foundations of Statistics for Data Science. In this module, you will learn the fundamental statistical concepts and techniques that are crucial for understanding and working with data in a data science context. Statistics form the backbone of data-driven decision-making and provide data scientists with the tools to analyze, interpret, and derive valuable insights from data.

We will begin by introducing descriptive statistics, which help summarize and describe the main features of a dataset. You'll learn about measures of central tendency, dispersion, and shape, and how these can provide a high-level understanding of your data. We'll discuss practical examples, such as analyzing customer demographics or product sales, to illustrate how descriptive statistics can be used to extract meaningful information.

Next, we will delve into probability theory, exploring the concepts of randomness and uncertainty that underpin statistical analysis. You will learn about various probability distributions and how they can model real-world phenomena, such as the likelihood of a specific event occurring or the distribution of customer spending. By understanding probability, you will be better equipped to make informed decisions and predictions based on data.

As we progress through the module, we will cover inferential statistics, which allow us to make generalizations about a population based on a sample. You'll learn about estimation techniques, confidence intervals, and the importance of understanding sampling variability. We'll use practical examples, such as predicting election results or estimating the average income in a city, to demonstrate the power of inferential statistics in data science applications.

Finally, we'll introduce hypothesis testing, a vital statistical tool that helps data scientists test assumptions, compare different groups, or evaluate the impact of an intervention. We'll discuss the logic behind hypothesis testing, the role of significance levels and p-values, and how to avoid common pitfalls. Through practical examples, such as A/B testing for marketing campaigns or analyzing the effectiveness of a new treatment, you will learn how to apply hypothesis testing to real-world data science problems.

By the end of this module, you will have a strong foundation in statistics, enabling you to confidently analyze and interpret data. This knowledge will serve as a crucial building block for the machine learning techniques covered in later modules, as well as for the broader field of data science.

Descriptive Statistics

Descriptive statistics provide a way to summarize and describe the main features of your dataset. They help you gain a high-level understanding of the data and identify patterns, trends, and potential outliers. In this section, we'll discuss the key measures of central tendency, dispersion, and shape, along with practical examples to help you grasp the concepts.

Measures of Central Tendency

- **Mean:** The mean, or average, is the sum of all data points divided by the total number of points. It's a common measure of central tendency and can be useful for understanding the overall distribution of the data. However, the mean can be sensitive to extreme values, which may not accurately represent the center of the data. For example, when analyzing the prices of houses in a neighborhood, a few very expensive houses may cause the mean to be higher than the typical house price.

- **Median:** The median is the middle value of a dataset when the data points are arranged in ascending order. If there is an odd number of data points, the median is the exact middle value; if there is an even number of points, the median is the average of the two middle values. The median is less sensitive to extreme values and can provide a more accurate representation of the center of the data. In the house price example, the median would be a better measure of central tendency, as it would not be influenced by a few exceptionally high-priced houses.

- **Mode:** The mode is the most frequently occurring value in a dataset. It can be useful for understanding the most common value in your data, particularly for categorical variables. For example, when analyzing the colors of cars in a parking lot, the mode would represent the most common car color.

Measures of Dispersion

- **Range:** The range is the difference between the highest and lowest values in a dataset. It provides a simple measure of the overall spread of the data. However, it can be sensitive to extreme values and may not accurately represent the dispersion of the data. For example, when analyzing the ages of people in a group, a single very old or very young person may cause the range to be larger than the typical age spread.

- **Variance:** The variance measures the average squared difference between each data point and the mean. It provides an indication of how spread out your data points are from the mean. A larger variance indicates that the data points are more spread out, while a smaller variance suggests that the data points are closer to the mean.

- **Standard Deviation:** The standard deviation is the square root of the variance. It is a commonly used measure of dispersion because it has the same units as the original data. A larger standard deviation indicates more variation in the data, while a smaller standard deviation indicates less variation.

Measures of Shape

- **Skewness:** Skewness measures the asymmetry of the distribution of your data. A positive skew indicates that the data is more spread out to the right of the mean, while a negative skew indicates that the data is more spread out to the left of the mean. For example, when analyzing the distribution of income in a population, you may find that the data is positively skewed, with a long tail of high-income earners.

- **Kurtosis:** Kurtosis measures the "tailedness" or the presence of outliers in your data. A high kurtosis indicates that the data has heavy tails and more outliers, while a low kurtosis indicates that the data has lighter tails and fewer outliers.

Understanding and applying descriptive statistics will allow you to better analyze your data, identify trends and patterns, and make informed decisions based on your findings.

Probability Theory

Probability theory is the branch of mathematics that deals with the analysis of random phenomena and the quantification of uncertainty. As a data scientist, understanding probability is crucial for making informed decisions and predictions based on your data. In this section, we'll discuss the fundamental concepts of probability theory and provide practical examples to illustrate their applications in data science.

- **Probability:** Probability is a measure of the likelihood that a particular event will occur, expressed as a number between 0 and 1. A probability of 0 means the event is impossible, while a probability of 1 means the event is certain. For example, the probability of flipping a fair coin and getting heads is 0.5, as there is an equal chance of getting heads or tails.

- **Random Variables:** A random variable is a variable whose value is determined by the outcome of a random event. There are two types of random variables: discrete and continuous. Discrete random variables take on a finite or countable number of possible values, such as the number of heads in a series of coin flips. Continuous random variables can take on an infinite number of values within a given interval, such as the height of a randomly selected person.

- **Probability Distributions:** A probability distribution is a function that describes the likelihood of each possible outcome for a random variable. For discrete random variables, the probability distribution is called a probability mass function (PMF), while for continuous random variables, it is called a probability density function (PDF). Some common probability distributions include the uniform, binomial, normal, and exponential distributions. Understanding different probability distributions can help you model real-world phenomena and make predictions based on your data.

Example: The normal distribution, also known as the Gaussian distribution, is a continuous probability distribution that is characterized by its bell-shaped curve. It is widely used in data science to model naturally occurring phenomena, such as human heights or test scores. By knowing the mean and standard deviation of a normal distribution, you can calculate the probability of a specific value or range of values occurring within the distribution.

- **Conditional Probability:** Conditional probability is the probability of an event occurring given that another event has already occurred. It is denoted as P(A|B), which represents the probability of event A happening, given that event B has occurred. Understanding conditional probability is essential for making predictions based on partial information and incorporating new evidence into your analysis.

  Example: Suppose you are analyzing the likelihood of customers making a purchase on your website. You might be interested in the probability of a customer making a purchase, given that they have added an item to their shopping cart. By calculating the conditional probability, you can estimate the effectiveness of your checkout process and make data-driven improvements.

- **Bayes' Theorem:** Bayes' theorem is a fundamental result in probability theory that relates the conditional probability of two events. It is particularly useful for updating prior beliefs or probabilities based on new evidence. Bayes' theorem is expressed as $P(A|B) = (P(B|A) * P(A)) / P(B)$, where P(A) and P(B) are the prior probabilities of events A and B, and P(B|A) is the likelihood of event B occurring given that event A has occurred.

Example: In a medical context, Bayes' theorem can be used to update the probability of a patient having a particular disease based on the results of a diagnostic test. By incorporating the sensitivity and specificity of the test, as well as the prior probability of the disease in the population, you can compute a more accurate estimate of the patient's disease status.

Understanding and applying probability theory will allow you to quantify uncertainty, model complex phenomena, and make informed decisions based on your data. This knowledge will serve as a foundation for the statistical techniques covered

Inferential Statistics

Inferential statistics allow you to make generalizations about a population based on a sample of data. By using inferential techniques, you can estimate population parameters, assess relationships between variables, and make predictions about future outcomes. In this section, we'll discuss key concepts and techniques in inferential statistics and provide practical examples to demonstrate their applications in data science.

- **Sampling:** Sampling is the process of selecting a subset of observations from a population to make inferences about the entire population. There are various sampling techniques, such as simple random sampling, stratified sampling, and cluster sampling, each with its advantages and limitations. The choice of sampling method depends on the research question, the structure of the population, and the resources available.

  Example: If you want to estimate the average income of residents in a city, you might use stratified sampling to ensure that your sample is representative of different income groups, neighborhoods, and demographics.

- **Estimation:** Estimation is the process of calculating a value, known as an estimator, to approximate a population parameter based on sample data. There are two main types of estimation: point estimation and interval estimation. Point estimation provides a single value as the best estimate of the population parameter, while interval estimation provides a range of values, known as a confidence interval, within which the population parameter is likely to fall.

  Example: Based on your stratified sample of city residents, you might estimate the average income using the sample mean as a point estimator, and calculate a 95% confidence interval to represent the range within which the true population mean is likely to lie.

- **Sampling Distribution:** The sampling distribution is the probability distribution of a given estimator when calculated from multiple random samples of the same size from the same population. Understanding the properties of the sampling distribution, such as its mean, variance, and shape, is crucial for making inferences about population parameters.

  Example: If you were to repeatedly draw samples from the city's population and calculate the sample mean for each, the distribution of these sample means would form the sampling distribution of the mean. The Central Limit Theorem states that, under certain conditions, the sampling distribution of the mean will be approximately normal, regardless of the shape of the population distribution.

- **Hypothesis Testing:** Hypothesis testing is a powerful inferential technique that allows you to test assumptions, compare different groups, or evaluate the impact of interventions based on your data. Hypothesis testing involves setting up a null hypothesis, which represents the status quo or no effect, and an alternative hypothesis, which represents the effect or relationship you want to test. You then calculate a test statistic and compare it to a critical value or compute a p-value to determine the strength of the evidence against the null hypothesis.

  Example: Suppose you want to test whether a new marketing campaign has increased sales for your online store. You could set up a null hypothesis stating that there is no difference in sales before and after the campaign, and an alternative hypothesis stating that sales have increased. By calculating a test statistic, such as the t-statistic, and comparing it to a critical value or computing a p-value, you can determine whether the observed increase in sales is statistically significant.

Mastering inferential statistics will allow you to make generalizations about populations, estimate parameters, and test hypotheses based on your data. This knowledge will enable you to draw meaningful conclusions and make data-driven decisions in a variety of data science applications.

Hypothesis Testing

Hypothesis testing is a fundamental method in inferential statistics that allows you to make decisions based on your data. It enables you to test claims, compare groups, and evaluate the effectiveness of interventions. In this section, we'll cover the steps involved in hypothesis testing and discuss various tests and their applications in data science.

- **Formulating Hypotheses:** The first step in hypothesis testing is to define the null hypothesis (H0) and the alternative hypothesis (H1). The null hypothesis typically represents a baseline or no effect scenario, while the alternative hypothesis represents the claim or effect you want to test.

  Example: If you want to determine whether a new drug is more effective than a standard treatment, the null hypothesis might be that there is no difference in effectiveness between the two treatments, and the alternative hypothesis might be that the new drug is more effective than the standard treatment.

- **Selecting a Test Statistic and Significance Level:** The test statistic is a numerical value calculated from your sample data, which is used to make a decision about the null hypothesis. The choice of test statistic depends on the type of data and the hypothesis being tested. Common test statistics include the t-statistic, z-statistic, and chi-square statistic. The significance level (α) is the probability of rejecting the null hypothesis when it is true, and is usually set at 0.05 or 0.01.

- **Calculating the Test Statistic and P-Value:** Once you have chosen a test statistic and significance level, you can calculate the test statistic from your sample data. The p-value is the probability of observing a test statistic as extreme or more extreme than the one calculated, assuming the null hypothesis is true. A smaller p-value indicates stronger evidence against the null hypothesis.

- **Making a Decision:** To make a decision about the null hypothesis, compare the p-value to the significance level. If the p-value is less than or equal to the significance level, you reject the null hypothesis in favor of the alternative hypothesis. If the p-value is greater than the significance level, you fail to reject the null hypothesis, meaning there is not enough evidence to support the alternative hypothesis.

  Example: Using the t-test to compare the effectiveness of the new drug and the standard treatment, you might obtain a p-value of 0.02. If your significance level is set at 0.05, you would reject the null hypothesis and conclude that there is evidence to suggest the new drug is more effective than the standard treatment.

Types of Hypothesis Tests

- **One-sample t-test:** The one-sample t-test is used to compare the mean of a single sample to a known population mean or a hypothesized value. This test is useful when you want to determine if your sample mean is significantly different from a specified value.

- **Two-sample t-test:** The two-sample t-test is used to compare the means of two independent samples. This test is useful when you want to determine if there is a significant difference between the means of two groups, such as a treatment group and a control group.

- **Paired t-test:** The paired t-test is used to compare the means of two related samples, such as measurements taken before and after an intervention on the same subjects. This test is useful when you want to determine if there is a significant change in the mean of a variable due to an intervention.

- **Chi-square test:** The chi-square test is used to compare observed frequencies in a contingency table to the frequencies expected under the assumption of independence between categorical variables. This test is useful when you want to determine if there is a significant association between two categorical variables.

Understanding hypothesis testing and its various applications will allow you to make informed decisions, test claims, and evaluate the effectiveness of interventions based on your data. Hypothesis testing is a crucial tool in data science, allowing you to draw meaningful conclusions and make data-driven decisions in a wide range of scenarios.

- **Analysis of Variance (ANOVA):** ANOVA is used to compare the means of three or more independent samples to determine if there is a significant difference among the group means. This test is useful when you want to analyze the effect of multiple factors or treatments on a continuous outcome variable.

- **Regression Analysis:** Regression analysis is used to explore the relationship between a dependent variable and one or more independent variables. Linear regression, for example, estimates the best-fitting straight line that describes the relationship between the variables. This technique is useful for predicting the value of a dependent variable based on the values of the independent variables.

  Example: Suppose you want to predict house prices based on factors such as square footage, number of bedrooms, and location. You can use multiple linear regression to build a model that estimates house prices based on these variables.

- **Non-parametric Tests:** Non-parametric tests are used when the assumptions of parametric tests, such as normality or homoscedasticity, are not met. Examples of non-parametric tests include the Wilcoxon rank-sum test, the Kruskal-Wallis test, and the Spearman's rank correlation coefficient. These tests are useful when you want to make inferences about your data without relying on strict assumptions about the underlying distribution.

Mastering hypothesis testing and its various applications will allow you to draw meaningful conclusions and make data-driven decisions in a wide range of data science scenarios. Hypothesis testing is an essential tool that enables you to test claims, compare groups, and evaluate the impact of interventions based on your data.

Bayesian Statistics

Bayesian statistics is an alternative approach to classical (frequentist) statistics that incorporates prior knowledge and updates beliefs based on observed data. In this section, we'll introduce you to the concepts and techniques in Bayesian statistics, including Bayes' theorem, Bayesian inference, and Bayesian model comparison. You'll learn how to apply Bayesian methods to real-world data science problems and see the advantages of this approach.

- **Bayes' Theorem:** Bayes' theorem is the foundation of Bayesian statistics. It provides a way to update the probability of an event or hypothesis (called the posterior probability) based on new evidence. The formula for Bayes' theorem is:

  P(A|B) = (P(B|A) * P(A)) / P(B)

  Here, P(A|B) is the posterior probability of event A given event B, P(B|A) is the likelihood of event B given event A, P(A) is the prior probability of event A, and P(B) is the probability of event B.

  Example: Suppose you have a test for a rare disease that has a 99% true positive rate and a 1% false positive rate. If the prevalence of the disease is 0.1% in the population, you can use Bayes' theorem to calculate the probability that a person who tests positive actually has the disease:

  P(Disease|Positive) = (P(Positive|Disease) * P(Disease)) / P(Positive)

- **Bayesian Inference:** Bayesian inference is the process of updating the prior probability based on observed data to obtain the posterior probability. In Bayesian inference, the prior represents your initial belief about a parameter or hypothesis, the likelihood quantifies how well the observed data supports the hypothesis, and the posterior represents the updated belief after considering the new evidence.

Example: In drug development, you can use Bayesian inference to update your belief about the effectiveness of a new treatment based on clinical trial data. You start with a prior belief about the treatment effect, update it with the observed data (e.g., patient outcomes), and obtain a posterior distribution that represents your updated belief about the treatment effect.

- **Bayesian Model Comparison:** Bayesian model comparison involves comparing different models or hypotheses to determine which one best explains the observed data. This can be done using measures such as the Bayes factor or the Deviance Information Criterion (DIC). The Bayes factor is the ratio of the marginal likelihoods of two models, and the DIC is a measure of model fit that takes into account model complexity.

  Example: Suppose you have two competing models to predict customer churn: one using only demographic features and another using both demographic and behavioral features. You can use Bayesian model comparison to determine which model better explains the observed data, helping you choose the best model for predicting churn.

Understanding and applying Bayesian statistics will allow you to leverage prior knowledge and make more robust decisions based on data. This approach is particularly useful in data science scenarios where data is scarce or noisy, or when incorporating domain knowledge is crucial for making accurate predictions and inferences.

Experimental Design

In this section, we will delve into experimental design, a crucial aspect of data science that allows you to plan and conduct experiments to obtain high-quality, reliable data. A well-designed experiment can help you establish causal relationships and evaluate the impact of interventions. We will discuss the essential principles of experimental design, such as randomization, blocking, control groups, and replication, and demonstrate how to apply these principles in real-world data science problems.

- **Randomization:** Randomization is the process of randomly allocating subjects or units to different treatment groups or conditions. This technique helps eliminate selection bias, ensures that treatment groups are comparable, and allows you to make valid inferences about the treatment effects.

  Example: When testing a new medication, you can randomly assign patients to either the treatment group (receiving the medication) or the control group (receiving a placebo). This ensures that any differences in outcomes between the groups can be attributed to the medication rather than confounding factors.

- **Blocking:** Blocking is a technique used to reduce the effects of confounding variables by grouping similar subjects or units together and randomizing within those groups. Blocking helps improve the precision of your estimates and reduces the impact of variability in the experiment.

  Example: In an agricultural experiment, you may want to test the effect of different fertilizers on crop yield. To account for variability in soil quality, you can block plots of land with similar soil characteristics and randomize the fertilizer treatments within each block.

- **Control Groups:** Control groups are used to provide a baseline against which the effects of a treatment can be compared. By including a control group, you can isolate the effect of the treatment and ensure that any observed differences are due to the treatment rather than other factors.

  Example: In a marketing experiment, you may want to test the effectiveness of a new promotional campaign. By including a control group that does not receive the campaign, you can compare the sales performance between the treatment and control groups to measure the impact of the campaign.

- **Replication:** Replication involves repeating an experiment multiple times to ensure that the results are consistent and reliable. Replication helps to account for variability and increases the precision of your estimates.

  Example: In a laboratory experiment, you may want to test the effect of a chemical compound on cell growth. By performing the experiment multiple times using different cell cultures, you can ensure that the observed effects are consistent and not due to random chance or errors in measurement.

Understanding and applying the principles of experimental design will allow you to plan and conduct experiments that yield reliable, valid results. This knowledge is essential for making data-driven decisions and evaluating the impact of interventions in various data science scenarios. With well-designed experiments, you can establish causal relationships and provide valuable insights that inform decision-making processes.

Power Analysis and Sample Size Determination

In this section, we'll explore power analysis and sample size determination, two essential concepts in experimental design and data science. Power analysis helps you determine the sample size required to detect an effect of a given size with a specified level of confidence. Understanding power analysis and sample size determination ensures that your study is adequately powered to detect meaningful effects, reducing the risk of false negatives (Type II errors) and increasing the reliability of your results.

- **Statistical Power:** Statistical power is the probability of correctly rejecting the null hypothesis when the alternative hypothesis is true. In other words, it's the probability of detecting an effect when one exists. Power is influenced by factors such as effect size, sample size, and significance level (alpha). A common target for statistical power is 0.8 or 80%, which means that there is an 80% chance of detecting an effect if it exists.

- **Effect Size:** Effect size is a quantitative measure of the magnitude of the difference between groups or the strength of a relationship between variables. Common effect size measures include Cohen's d, correlation coefficient (r), and odds ratio. Effect size is a crucial component of power analysis, as larger effect sizes are easier to detect with smaller sample sizes.

- **Sample Size Determination:** To determine the required sample size for your study, you'll need to consider the desired power, effect size, and significance level. Various methods and software tools can help you calculate the necessary sample size, such as power analysis calculators, simulation studies, or analytical formulas.

Example: Suppose you want to conduct a study to evaluate the effectiveness of a new teaching method. Based on previous research, you expect a moderate effect size (Cohen's d = 0.5). You decide to use a significance level of 0.05 and aim for a power of 0.8. Using a power analysis calculator, you determine that you need a total sample size of 128 participants (64 in each group) to achieve the desired power.

Conducting a power analysis and determining the appropriate sample size will allow you to ensure that your study is adequately powered to detect meaningful effects. This increases the reliability of your results and allows you to make more accurate inferences about the population. In data science, understanding power analysis and sample size determination is crucial for designing robust experiments and making data-driven decisions based on reliable evidence.

Multiple Comparison Problem and Corrections

In this section, we will explore the multiple comparison problem and the various methods to correct for it. The multiple comparison problem arises when you perform several statistical tests simultaneously, increasing the likelihood of false positives (Type I errors). By understanding and applying multiple comparison corrections, you can control the false positive rate and increase the reliability of your results in data science projects.

- **Familywise Error Rate (FWER):** FWER is the probability of making at least one false positive among all the comparisons you perform. The goal of multiple comparison corrections is to control the FWER at a pre-specified level, such as 0.05, to maintain the overall error rate.

- **Bonferroni Correction:** The Bonferroni correction is a simple and conservative method for adjusting the significance level (alpha) when performing multiple tests. To apply the Bonferroni correction, divide the desired overall significance level by the number of comparisons (m). The resulting value is the adjusted significance level for each individual test.

  Example: Suppose you are conducting an experiment with three treatment groups (A, B, and C) and wish to compare the mean outcomes for each pair of groups. You need to perform three comparisons (A vs. B, A vs. C, and B vs. C). With the Bonferroni correction, the adjusted significance level for each comparison is $0.05/3 \approx 0.0167$.

- **Holm-Bonferroni Correction:** The Holm-Bonferroni correction is a stepwise method that is less conservative than the Bonferroni correction. To apply the Holm-Bonferroni correction, order the p-values from the multiple tests in ascending order and compare each p-value to its corresponding adjusted significance level (i/m * alpha), where i is the rank of the p-value, m is the number of comparisons, and alpha is the desired overall significance level. Stop when the first non-significant p-value is encountered.

- **False Discovery Rate (FDR) Control:** FDR is the expected proportion of false positives among the significant results. FDR control methods, such as the Benjamini-Hochberg procedure, provide a less conservative approach to multiple comparison corrections. To apply the Benjamini-Hochberg procedure, order the p-values in ascending order and compare each p-value to its corresponding FDR threshold (i/m * q), where i is the rank of the p-value, m is the number of comparisons, and q is the desired FDR level. The largest p-value that meets this criterion and all smaller p-values are considered significant.

Understanding the multiple comparison problem and applying appropriate corrections will allow you to control the false positive rate and increase the reliability of your results in data science projects. These techniques are essential when dealing with multiple hypotheses, large datasets, or complex experimental designs, ensuring that your conclusions are based on sound statistical principles.

Non-linear Regression

In this section, we will focus on non-linear regression, a powerful and flexible statistical method that models the relationship between a dependent variable and one or more independent variables when the relationship is not strictly linear. Non-linear regression allows you to fit complex, non-linear functions to your data, providing more accurate predictions and insights in various data science scenarios.

- **Non-linear Functions:** Non-linear functions are mathematical functions that do not follow a straight line when plotted on a graph. They can represent various types of relationships between variables, such as exponential, logarithmic, polynomial, or sigmoid functions.

- **Fitting Non-linear Models:** Fitting a non-linear model involves finding the function that best describes the relationship between the dependent and independent variables by minimizing the residuals (the differences between the observed and predicted values). Non-linear regression typically requires iterative optimization techniques, such as gradient descent, Gauss-Newton, or Levenberg-Marquardt algorithms, to estimate the model parameters.

- **Model Selection:** When working with non-linear regression, it is essential to select the appropriate function to model the relationship between the variables. Model selection can be based on theoretical considerations, prior knowledge, or visual inspection of the data. Techniques such as cross-validation, Akaike Information Criterion (AIC), or Bayesian Information Criterion (BIC) can help you compare the performance of different models and choose the most appropriate one.

Example: Suppose you are analyzing the growth of a bacterial population over time. Based on biological knowledge, you know that bacterial growth often follows a logistic growth pattern. You can use non-linear regression to fit a logistic growth model to your data, with the dependent variable being the bacterial population size and the independent variable being time. The logistic growth model will allow you to estimate important parameters, such as the maximum population size and the growth rate, providing valuable insights into the dynamics of the bacterial population.

- **Assessing Model Fit:** After fitting a non-linear model, it is crucial to assess the quality of the fit. You can use various diagnostic tools and goodness-of-fit measures, such as residual plots, R-squared, or root mean squared error (RMSE), to evaluate the adequacy of the model and identify potential issues, such as heteroscedasticity or non-normality of residuals.

Understanding non-linear regression and its applications will allow you to model complex relationships in your data that are not adequately captured by linear models. Non-linear regression is a versatile and powerful tool in data science, allowing you to obtain more accurate predictions and insights in various fields, such as biology, finance, engineering, and social sciences.

Time Series Analysis

In this section, we will delve into time series analysis, a vital statistical method for analyzing data collected over time. Time series analysis allows you to uncover underlying patterns, trends, and seasonality, as well as make forecasts and predictions. This technique is invaluable in various data science applications, such as finance, economics, weather forecasting, and sales.

Components of Time Series

A time series consists of four main components, which can be used to describe the patterns and structure of the data:

- **Trend:** The long-term movement or direction of the data.
- **Seasonality:** The recurring patterns or cycles that occur at regular intervals.
- **Cyclical Variation:** Fluctuations that are not seasonal but follow a certain pattern.
- **Irregular Variation:** Random fluctuations or noise that cannot be attributed to any specific pattern.

Autocorrelation

Autocorrelation is a measure of the correlation between a time series and its lagged values. It is a crucial concept in time series analysis, as it can help identify the presence of trends, seasonality, or dependence between observations.

Stationarity

A stationary time series has constant mean, variance, and autocorrelation structure over time. Many time series analysis techniques require the data to be stationary. If your data is non-stationary, you may need to apply transformations, such as differencing or detrending, to achieve stationarity.

Time Series Forecasting Models

Various models can be used to forecast future values of a time series based on its historical data. Some common time series forecasting models include:

- **Autoregressive (AR) Models:** These models assume that the current value of the series depends linearly on its previous values.
- **Moving Average (MA) Models:** These models use a linear combination of past errors to predict future values.
- **Autoregressive Integrated Moving Average (ARIMA) Models:** These models combine the AR and MA components and include an integration step for handling non-stationary data.
- **Seasonal Decomposition of Time Series (STL) and Exponential Smoothing State Space Model (ETS):** These models account for trends and seasonality in the data.
- **Prophet:** A procedure developed by Facebook for forecasting time series data with multiple seasonalities and holidays.
- Example: Suppose you are a data scientist at an e-commerce company, and you want to forecast the monthly sales for the next year. By applying time series analysis techniques, such as decomposition or ARIMA modeling, you can identify the underlying patterns in the historical sales data, estimate the model parameters, and make predictions for future sales. This information can help your company make informed decisions about inventory management, marketing campaigns, and resource allocation.

Understanding time series analysis and its applications will allow you to unlock valuable insights from temporal data and make accurate predictions for a wide range of data science scenarios. Time series analysis is an essential tool for data scientists working in fields that involve time-dependent data, enabling informed decision-making and effective planning.

Module 4 Exam Questions

1. What is the primary purpose of descriptive statistics?

a) To make predictions

b) To test hypotheses

c) To summarize and describe the main features of a dataset

d) To model complex relationships

2. Which of the following is NOT a measure of central tendency?

a) Mean

b) Median

c) Mode

d) Range

3. Which probability distribution is commonly used to model the number of successes in a fixed number of Bernoulli trials?

a) Normal distribution

b) Poisson distribution

c) Binomial distribution

d) Uniform distribution

4. In hypothesis testing, a Type I error occurs when:

a) A true null hypothesis is rejected

b) A false null hypothesis is rejected

c) A true null hypothesis is not rejected

d) A false null hypothesis is not rejected

5. Which method is used to control the Familywise Error Rate (FWER)?

a) Bonferroni correction

b) Benjamini-Hochberg procedure

c) Holm-Bonferroni correction

d) None of the above

6. Which of the following is a non-linear function?

a) Linear regression

b) Exponential function

c) Straight line

d) All of the above

7. Which of the following techniques is used to assess the quality of a non-linear regression model?

a) Residual plots

b) Cross-validation

c) Autocorrelation

d) Detrending

8. What is the main goal of time series analysis?

a) To uncover underlying patterns and make forecasts

b) To fit complex functions to data

c) To control the false positive rate

d) To minimize residuals

9. Which of the following is a characteristic of a stationary time series?

a) Constant mean over time

b) Increasing variance over time

c) Decreasing autocorrelation over time

d) None of the above

10. Which time series forecasting model combines autoregressive and moving average components?

a) ARIMA

b) ETS

c) Prophet

d) STL

11. Which of the following is an example of supervised machine learning?

a) Regression

b) Clustering

c) Dimensionality reduction

d) Reinforcement learning

12. Which Python library is primarily used for data manipulation and analysis?

a) NumPy

b) Pandas

c) Matplotlib

d) Seaborn

13. Which plot is commonly used to visualize the distribution of a continuous variable?

a) Bar plot

b) Histogram

c) Scatter plot

d) Box plot

14. What is the purpose of feature engineering in data preprocessing?

a) To clean the data

b) To transform raw data into more useful features for modeling

c) To visualize the data

d) To collect and import data

15. Which of the following is an example of unsupervised machine learning?

a) Classification

b) Regression

c) Clustering

d) None of the above

16. Which type of plot is particularly useful for visualizing the relationship between two continuous variables?

a) Bar plot

b) Pie chart

c) Histogram

d) Scatter plot

17. What is the main focus of the module "Foundations of Statistics for Data Science"?

a) Python programming

b) Data visualization

c) Statistical techniques and concepts

d) Machine learning

18. Which of the following is NOT an example of a deep learning model?

a) Convolutional Neural Networks

b) Recurrent Neural Networks

c) Linear Regression

d) None of the above

19. In the context of data science, what does EDA stand for?

a) Error Detection and Analysis

b) Exploratory Data Analysis

c) Effective Data Aggregation

d) Essential Data Algorithms

20. What is the primary goal of a data scientist when performing exploratory data analysis (EDA)?

a) To fit the best machine learning model

b) To gain insights and identify patterns in the data

c) To present the final results to stakeholders

d) To collect and store the data

Answers:

1. c
2. d
3. c
4. a
5. a
6. b
7. a
8. a
9. a
10. a
11. a
12. b
13. b
14. b
15. c
16. d
17. c
18. c
19. b
20. b
21.

Module 5. Supervised Machine Learning

Introduction to Supervised Machine Learning

Welcome to Module 5 on Supervised Machine Learning. In this module, you will learn the fundamental concepts and techniques of supervised machine learning, one of the most widely used branches of machine learning. Supervised learning algorithms rely on labeled data, where each data point is associated with a known output or target. The goal of these algorithms is to learn the underlying relationships between input features and output labels so that they can make predictions on new, unseen data.

Throughout this module, we will delve into the theory and practical applications of supervised machine learning, illustrating its importance and versatility in solving real-world problems. You will be introduced to various algorithms, ranging from simple linear regression to more complex techniques like support vector machines and decision trees. By understanding the strengths and weaknesses of each method, you will be better equipped to choose the most suitable approach for a given task.

To demonstrate the power of supervised machine learning, we will explore practical examples from various industries, such as healthcare, finance, and marketing. For instance, we will look at how banks can leverage credit scoring models to predict the likelihood of a customer defaulting on a loan, or how e-commerce companies can use customer data to create personalized product recommendations.

In addition to learning the theory behind these algorithms, you will also gain hands-on experience in implementing them using popular Python libraries, such as scikit-learn. This will enable you to not only understand the inner workings of these models but also appreciate the practical challenges and considerations involved in applying them to real-world datasets.

By the end of this module, you will have a solid foundation in supervised machine learning, which will empower you to harness the predictive power of data and make informed decisions in your professional or personal projects.

Regression Algorithms

In this section, we will explore regression algorithms, a type of supervised machine learning technique used to predict continuous target variables. These algorithms are designed to help you understand the relationship between input features (independent variables) and a continuous output (dependent variable) by modeling the underlying function that connects them. Here, we will discuss some of the most commonly used regression algorithms, their key principles, and practical examples to guide your understanding.

Linear Regression

Linear regression is a simple yet powerful algorithm that models the relationship between input features and a continuous target variable as a linear combination of the input variables. The algorithm tries to find the best-fitting line, which minimizes the sum of the squared differences between the observed and predicted values. This method is particularly useful when the relationship between the variables is approximately linear.

Example: Predicting house prices based on factors such as square footage, number of bedrooms, and location. You can use linear regression to create a model that predicts the price of a house based on these features.

Multiple Linear Regression

Multiple linear regression is an extension of linear regression that allows you to include multiple input features. This algorithm also models the relationship between input variables and a continuous target variable as a linear combination of the input variables. It helps you capture the combined effect of multiple factors on the target variable.

Example: Expanding on the house price prediction example, you can use multiple linear regression to include additional factors such as the age of the house, proximity to schools, and local crime rates.

Polynomial Regression

Polynomial regression is a technique used when the relationship between input features and the target variable is non-linear. This algorithm models the relationship as a polynomial function, allowing you to capture more complex patterns in the data.

Example: Predicting the trajectory of a projectile based on initial velocity, angle, and air resistance. Using polynomial regression, you can model the non-linear relationship between these factors and the projectile's distance traveled.

Ridge Regression

Ridge regression is a variant of linear regression that introduces a regularization term, which helps reduce the complexity of the model and prevent overfitting. This algorithm is particularly useful when you have a high number of input features or when the features are highly correlated.

Example: Predicting stock prices based on a large number of financial indicators. You can use ridge regression to create a model that considers all these factors without overfitting the training data.

Lasso Regression

Lasso (Least Absolute Shrinkage and Selection Operator) regression is another variant of linear regression that introduces a different regularization term. Lasso regression not only helps prevent overfitting but also performs feature selection by driving some feature coefficients to zero, effectively excluding them from the model.

Example: Predicting patient outcomes in healthcare based on a large set of medical features. You can use Lasso regression to identify the most relevant features and create a more parsimonious model.

As you progress through this section, you will gain a deeper understanding of these regression algorithms, their assumptions, and limitations. By mastering these techniques, you will be well-equipped to tackle a wide range of prediction tasks involving continuous target variables in various domains.

Classification Algorithms

In this section, we will delve into classification algorithms, another type of supervised machine learning technique used to predict categorical target variables. These algorithms aim to classify data points into distinct categories or classes based on their input features. We will discuss some of the most widely used classification algorithms, their key principles, and practical examples to enrich your understanding.

Logistic Regression

Logistic regression is a classification algorithm that models the probability of an input data point belonging to a specific category. It is particularly useful for binary classification problems, where the target variable has only two possible classes. The algorithm uses the logistic function to model the probability and applies a threshold to make the final classification.

Example: Predicting whether a customer will make a purchase based on factors such as age, income, and browsing history. You can use logistic regression to create a model that estimates the probability of a purchase and classifies customers accordingly.

Decision Trees

Decision trees are a type of classification algorithm that recursively split the input data into subsets based on the values of the input features. The algorithm builds a tree-like structure, where each internal node represents a feature and each leaf node represents a class label. Decision trees are easy to interpret and can capture complex relationships between features.

Example: Predicting whether a loan applicant is likely to default based on factors such as credit score, income, and loan amount. You can use a decision tree to create a model that represents the decision-making process and classifies applicants based on their risk.

Random Forest

Random forest is an ensemble learning method that combines multiple decision trees to make a more accurate and robust classification. The algorithm creates a diverse set of trees by using a random subset of input features and data points for each tree. The final classification is determined by aggregating the predictions from all the individual trees.

Example: Diagnosing diseases based on a set of symptoms and medical test results. You can use a random forest to create a model that considers multiple possible decision paths and achieves higher accuracy in predicting the correct diagnosis.

Support Vector Machines (SVM)

Support vector machines are a classification algorithm that aims to find the best separating hyperplane between classes in the input feature space. The algorithm maximizes the margin between the hyperplane and the closest data points from each class, known as support vectors. SVM can handle both linear and non-linear classification problems using the kernel trick.

Example: Classifying handwritten digits based on pixel values. You can use SVM to create a model that distinguishes between different digit classes even when the relationships between pixel values are complex and non-linear.

K-Nearest Neighbors (KNN)

K-nearest neighbors is a simple, instance-based classification algorithm that classifies a data point based on the majority class of its k nearest neighbors in the input feature space. The algorithm is non-parametric, which means it does not make any assumptions about the underlying distribution of the data.

Example: Predicting customer churn based on their recent activity and purchase patterns. You can use KNN to create a model that compares a customer to others with similar behavior and classifies them as likely or unlikely to churn.

As you progress through this section, you will develop a comprehensive understanding of these classification algorithms, their strengths and weaknesses, and when to use them in practice. By mastering these techniques, you will be well-prepared to tackle a wide range of classification tasks across various domains.

Model Evaluation and Validation

In this section, we will discuss model evaluation and validation, which are essential steps in the supervised machine learning process. Evaluating and validating your models helps you determine their performance, identify potential issues, and select the best model for your specific task. We will cover various evaluation metrics, methods, and techniques that will guide you through this critical phase of the machine learning pipeline.

Evaluation Metrics

Evaluation metrics are quantitative measures used to assess the performance of your machine learning models. Different metrics are suitable for different types of tasks, such as regression or classification. Here are some commonly used evaluation metrics:

- **Regression:** Mean Absolute Error (MAE), Mean Squared Error (MSE), and R-squared

- **Classification:** Accuracy, Precision, Recall, F1-score, and Area Under the Receiver Operating Characteristic (ROC) Curve (AUC-ROC)

Example: When predicting customer churn, you might use accuracy, precision, and recall to evaluate the performance of your classification model, as these metrics will help you understand how well your model identifies true positives and negatives.

Training and Testing Sets

To evaluate and validate your models, it's essential to split your data into separate training and testing sets. The training set is used to build the model, while the testing set is used to evaluate its performance on unseen data. This helps you estimate how well your model will generalize to new data points.

Example: When creating a model to predict house prices, you can use 70% of your data for training and 30% for testing, ensuring that the model's performance is assessed on data it has not seen before.

Cross-Validation

Cross-validation is a technique used to estimate the performance of a model by dividing the data into multiple folds, training and testing the model on each fold, and averaging the results. This method helps reduce the impact of sampling bias and provides a more reliable performance estimate.

Example: When building a model to diagnose diseases, you can use k-fold cross-validation, where the data is divided into k equal-sized folds. The model is then trained on k-1 folds and tested on the remaining fold, repeating this process k times and averaging the evaluation metrics.

Model Selection and Hyperparameter Tuning

Model selection involves choosing the best model from a set of candidate models, while hyperparameter tuning focuses on finding the optimal configuration for a given model. Both steps are crucial in optimizing model performance.

Example: When classifying handwritten digits, you might try different classification algorithms, such as SVM, KNN, or random forest, and compare their performance using cross-validation. Once you've chosen the best algorithm, you can fine-tune its hyperparameters, such as the number of trees in a random forest or the value of k in KNN, to further improve its performance.

Model Validation

Model validation is the final step in evaluating your machine learning models. Once you have selected the best model and tuned its hyperparameters, you can validate its performance using the testing set, providing a final estimate of how well it will perform on new, unseen data.

Example: After selecting the best model for predicting stock prices and tuning its hyperparameters, you can use the testing set to validate its performance, giving you confidence in its ability to make accurate predictions on future stock prices.

As you work through this section, you will develop a comprehensive understanding of model evaluation and validation techniques, enabling you to assess, compare, and optimize your machine learning models effectively. Mastering these skills is crucial for building reliable and high-performing models that can be confidently deployed in real-world applications.

Feature Selection Techniques

In this section, you'll learn about feature selection techniques, which play a crucial role in building effective machine learning models. Feature selection is the process of choosing the most important features from the available dataset to train a model. It helps in reducing the complexity of the model, improving its performance, and minimizing the risk of overfitting. We'll explore several methods for feature selection, with practical examples to help you understand their application.

Filter Methods

Filter methods evaluate the relevance of each feature based on their individual characteristics, without taking the model into consideration. These methods are fast and straightforward, and can be applied before training the model.

- **Correlation Coefficient:** You can use correlation coefficients, such as Pearson's or Spearman's, to measure the linear relationship between a feature and the target variable. Features with high correlation are more likely to be important for the model.

- **Mutual Information:** This technique measures the dependence between two variables, capturing both linear and nonlinear relationships.

Wrapper Methods

Wrapper methods use a specific machine learning model to assess the relevance of features. They involve a search process, exploring different feature subsets to find the best one for the given model.

- **Recursive Feature Elimination (RFE):** RFE is an iterative method that starts with all features and eliminates the least important ones step by step.

- **Forward Selection & Backward Elimination:** In forward selection, you start with an empty set of features and add the most important one at each step. In backward elimination, you start with all features and remove the least important one at each step.

Embedded Methods

These methods perform feature selection as part of the model training process. They are specific to certain algorithms and often involve regularization techniques.

- **LASSO Regression:** LASSO (Least Absolute Shrinkage and Selection Operator) is a linear regression technique that adds an L1 penalty to the loss function, effectively shrinking some coefficients to zero and thus eliminating the corresponding features.

- **Decision Trees and Random Forests:** These tree-based methods inherently perform feature selection by splitting the data based on the most important features at each step.

Using these feature selection techniques will allow you to identify the most relevant features for your machine learning model, improve its performance, and reduce the risk of overfitting. When applying these methods, be sure to consider the characteristics of your dataset and the specific requirements of your model.

Dealing with Imbalanced Data

In this section, you'll learn about dealing with imbalanced data, which is a common challenge in real-world machine learning problems. Imbalanced data occurs when the distribution of classes in your target variable is significantly skewed, leading to an unequal representation of different classes. This can cause your model to be biased towards the majority class and underperform on the minority class. We'll explore several techniques to handle imbalanced data, with practical examples to help you understand their application.

Resampling Techniques

You can balance the class distribution by either oversampling the minority class, undersampling the majority class, or a combination of both.

- **Random Oversampling:** In this technique, you duplicate instances from the minority class to balance the class distribution.
- **Random Undersampling:** This method removes instances from the majority class to balance the class distribution.
- **SMOTE:** Synthetic Minority Over-sampling Technique (SMOTE) generates synthetic instances for the minority class using interpolation.

Adjusting Model Parameters

Certain algorithms, like decision trees and logistic regression, allow you to adjust class weights to give more importance to the minority class during training.

Using Different Evaluation Metrics

Accuracy might not be the best metric for imbalanced data. Instead, consider using other metrics, such as precision, recall, F1-score, or the area under the Receiver Operating Characteristic (ROC) curve.

Employing Specialized Algorithms

Some machine learning algorithms are specifically designed to handle imbalanced data, such as the BalancedRandomForestClassifier and EasyEnsembleClassifier from the imbalanced-learn library.

By using these techniques, you can effectively handle imbalanced data and build more robust machine learning models that perform well on both majority and minority classes. It's important to experiment with different approaches to find the best strategy for your specific dataset and problem.

Combining Techniques: You can also combine multiple techniques mentioned above to achieve even better performance. For instance, you can use resampling techniques along with adjusting model parameters or trying specialized algorithms. Always ensure to validate your model using appropriate evaluation metrics and cross-validation techniques to get a better understanding of its performance.

As you deal with imbalanced data, remember that there is no one-size-fits-all solution. Be prepared to try different techniques, fine-tune their parameters, and evaluate their performance using suitable metrics to find the best approach for your specific problem. With practice and experience, you'll become more adept at identifying the most effective techniques for handling imbalanced data in your machine learning projects.

Advanced Ensemble Techniques

In this section, you'll learn about advanced ensemble techniques, which are powerful methods for improving the performance of your machine learning models. Ensemble techniques combine the predictions of multiple base models to create a more accurate and robust final model. We'll explore several advanced ensemble techniques and provide practical examples to help you understand their application.

Bagging (Bootstrap Aggregating)

Bagging is an ensemble method that involves training multiple base models, usually decision trees, on different random subsets of the dataset. The subsets are created using bootstrapping, which involves sampling with replacement. The final prediction is obtained by averaging the predictions of all base models for regression tasks or by taking a majority vote for classification tasks.

Boosting

Boosting is another ensemble technique that combines several weak learners to create a strong learner. Unlike bagging, boosting trains base models sequentially, with each model learning from the errors of the previous model. Popular boosting algorithms include AdaBoost, Gradient Boosting, and XGBoost.Stacking

Stacking is a technique that involves training multiple base models on the dataset and then using their predictions as inputs for a meta-model, which makes the final prediction. The idea is that the meta-model can learn to leverage the strengths of individual base models, resulting in improved performance.

Applying these advanced ensemble techniques will allow you to significantly improve the performance of your machine learning models. Keep in mind that ensembling can sometimes increase the complexity of your model, so always evaluate the trade-offs between complexity and performance. Experiment with different combinations of base models and ensemble techniques to find the best approach for your specific problem.

Hyperparameter Tuning

In this section, you'll learn about hyperparameter tuning, a crucial aspect of building high-performing machine learning models. Hyperparameters are the configuration parameters of a model that are not learned during training. They significantly impact a model's performance, and finding the best combination of hyperparameters can lead to significant improvements in accuracy and generalization.

There are several techniques for hyperparameter tuning, and we'll explore some of the most popular ones along with practical examples:

Grid Search

Grid search is a methodical approach to hyperparameter tuning that involves evaluating all possible combinations of hyperparameter values. This exhaustive search can be computationally expensive, but it guarantees finding the best combination of hyperparameters for the given search space.

Random Search

Random search is a less exhaustive alternative to grid search that samples hyperparameter values from a specified distribution. This approach can be more efficient than grid search, especially when dealing with a large search space.

Bayesian Optimization

Bayesian optimization is an advanced technique that uses a probabilistic model to guide the search for optimal hyperparameters. It is more efficient than grid search and random search, especially for high-dimensional search spaces.

Applying these hyperparameter tuning techniques will allow you to significantly improve the performance of your supervised machine learning models. Keep in mind that tuning hyperparameters can be computationally expensive, especially for large search spaces or complex models. It's essential to balance the search for optimal hyperparameters with the available computational resources and the time you have for model development. Always start with a reasonable set of default hyperparameters and only fine-tune them if necessary. This approach will help you avoid spending excessive time on hyperparameter tuning and allow you to focus on other aspects of the machine learning pipeline, such as feature engineering and model evaluation.

Module 5 Exam Questions

1. What is the primary goal of supervised machine learning?

a) Clustering data points

b) Reducing dimensionality

c) Predicting outcomes based on input features

d) Finding hidden patterns in data

2. Which of these is not a regression algorithm?

a) Linear Regression

b) Logistic Regression

c) Lasso Regression

d) Ridge Regression

3. Which evaluation metric is commonly used for classification tasks?

a) Mean Absolute Error

b) Mean Squared Error

c) F1-score

d) R-squared

4. In k-fold cross-validation, how is the data divided?

a) Into k equal-sized folds

b) Into k-1 equal-sized folds

c) Into 70% training and 30% testing sets

d) Randomly without any specific ratio

5. What does the R-squared value represent in regression tasks?

a) The proportion of variance in the target variable explained by the model

b) The difference between the predicted and actual values

c) The sum of squared residuals

d) The average of the absolute errors

6. Which classification algorithm is based on the concept of distance between data points?

a) Random Forest

b) Decision Tree

c) Support Vector Machines

d) K-Nearest Neighbors

7. What does the term "overfitting" refer to in machine learning?

a) A model that performs poorly on both training and testing sets

b) A model that performs well on the training set but poorly on the testing set

c) A model that performs well on the testing set but poorly on the training set

d) A model that performs well on both training and testing sets

8. What is the purpose of splitting data into training and testing sets?

a) To avoid overfitting and evaluate the model's performance on unseen data

b) To reduce the dimensionality of the dataset

c) To merge different datasets

d) To apply data preprocessing techniques

9. Which Python library is commonly used for implementing machine learning algorithms?

a) NumPy

b) Pandas

c) Matplotlib

d) Scikit-learn

10. What is the primary goal of classification tasks in supervised machine learning?

a) Predicting continuous values

b) Predicting discrete classes or categories

c) Grouping similar data points

d) Reducing the number of features in a dataset

11. Which of the following is not an evaluation metric for regression tasks?

a) Mean Absolute Error

b) Mean Squared Error

c) Precision

d) R-squared

12. In the context of machine learning, what does "validation" mean?

a) Selecting the best model from a set of candidate models

b) Estimating the performance of a model using the testing set

c) Tuning the hyperparameters of a model

d) Calculating the evaluation metrics for a model

13. What is a hyperparameter in machine learning?

a) A parameter that is learned by the model during training

b) A parameter that is set before training the model

c) A feature used to train the model

d) A metric used to evaluate the model's performance

14. Which algorithm can be used for both classification and regression tasks?

a) Linear Regression

b) Support Vector Machines

c) Logistic Regression

d) K-Means

15. Which of the following is a dimensionality reduction technique?

a) Principal Component Analysis (PCA)

b) K-Nearest Neighbors

c) Decision Trees

d) Linear Regression

16. In a decision tree, what is a leaf node?

a) A node that has no children

b) A node that has exactly one child

c) A node that has two children

d) The root node of the tree

17. Which of the following is an ensemble learning technique?

a) Support Vector Machines

b) Random Forest

c) K-Nearest Neighbors

d) Logistic Regression

18. What is regularization in machine learning?

a) A technique to reduce overfitting by adding a penalty term to the loss function

b) A method to divide the dataset into training and testing sets

c) A technique to find the best hyperparameters for a model

d) A method to preprocess data and remove missing values

19. Which Python library is commonly used for data manipulation and preprocessing in data science?

a) NumPy

b) Pandas

c) Matplotlib

d) Scikit-learn

20. What is the purpose of feature engineering in machine learning?

a) To reduce overfitting by selecting the best features

b) To create new features or modify existing features to improve model performance

c) To visualize the features and their relationships

d) To split the dataset into training and testing sets

Answers:

1. c
2. b
3. c
4. a
5. a
6. d
7. b
8. a
9. d
10. b
11. c
12. a
13. b
14. b
15. a
16. a
17. b
18. a
19. b
20. b

Module 6. Unsupervised Machine Learning

Introduction to Unsupervised Machine Learning

Welcome to Module 6 on Unsupervised Machine Learning. In this module, you will delve into the world of unsupervised learning, one of the fundamental branches of machine learning techniques. Unlike supervised learning, unsupervised learning does not rely on labeled data to make predictions or uncover patterns. Instead, it focuses on identifying underlying structures and relationships within the data itself.

Throughout this module, we will explore various unsupervised learning algorithms and techniques that can help you unlock valuable insights from complex datasets. You will learn how unsupervised learning methods can enable you to discover hidden patterns, groupings, and relationships in your data that may not be immediately apparent or easily discernible using traditional analysis methods.

To illustrate the power and practical applications of unsupervised learning, let's consider a real-world example. Imagine a retail company that wants to improve its marketing efforts by better understanding its customers' purchasing behavior. The company has a vast dataset containing transactional data, but it is not labeled with customer segments or preferences. Unsupervised learning techniques, such as clustering algorithms, can help analyze this data and identify distinct groups of customers with similar purchasing habits, allowing the company to tailor its marketing campaigns and offers more effectively.

Another example is in the field of natural language processing (NLP), where unsupervised learning can be employed to analyze large collections of text documents. By leveraging techniques such as topic modeling or word embeddings, researchers and data scientists can identify common themes, trends, and relationships within the text data, enabling a deeper understanding of the content and its underlying structure.

As we progress through this module, you will learn about various unsupervised learning techniques, including clustering algorithms and dimensionality reduction methods. These powerful tools will equip you with the knowledge and skills to tackle complex data analysis tasks and uncover valuable insights hidden within your data. So let's dive in and discover the exciting world of unsupervised machine learning!

Clustering Algorithms

In this section, we will explore clustering algorithms, one of the most widely-used unsupervised learning techniques. Clustering aims to group similar data points together based on their features, while keeping dissimilar data points separate. This process helps you uncover hidden patterns and structures in your dataset, allowing for better decision-making and insights.

There are several popular clustering algorithms that you can apply to your data, each with its own strengths and weaknesses. We will discuss three of the most common algorithms: K-Means, Hierarchical Clustering, and DBSCAN.

K-Means Clustering

K-Means is a centroid-based clustering algorithm that seeks to partition your dataset into K distinct clusters. The algorithm works by iteratively updating the cluster centroids (the center of each cluster) until convergence is reached. Here are the main steps of the K-Means algorithm:

1. Initialize K random centroids in the feature space.
2. Assign each data point to the nearest centroid.
3. Update the centroids by calculating the mean of all data points within each cluster.
4. Repeat steps 2 and 3 until the centroids' positions no longer change significantly, or a predefined number of iterations is reached.

One challenge with K-Means is determining the optimal number of clusters (K). You can use techniques like the Elbow Method or Silhouette Analysis to help you choose an appropriate value for K.

Hierarchical Clustering

Hierarchical Clustering, as the name suggests, builds a tree-like structure to represent the hierarchical relationships between data points. There are two main approaches to hierarchical clustering: agglomerative (bottom-up) and divisive (top-down).

In agglomerative clustering, each data point initially forms its own cluster. The algorithm then iteratively merges the closest pairs of clusters until a single cluster remains. Conversely, divisive clustering starts with a single cluster containing all data points and recursively splits the clusters until each data point is in its own cluster.

You can visualize the clustering process using a dendrogram, which is a tree diagram that displays the hierarchical relationships between clusters. By cutting the dendrogram at a specific height, you can choose the desired number of clusters.

DBSCAN (Density-Based Spatial Clustering of Applications with Noise)

DBSCAN is a density-based clustering algorithm that groups data points based on their proximity and density. Unlike K-Means and hierarchical clustering, DBSCAN does not require you to specify the number of clusters beforehand. Instead, it can automatically identify clusters of varying shapes and sizes, as well as detect and handle noise.

The main idea behind DBSCAN is that a cluster is a dense region of data points separated by areas of lower point density. The algorithm requires two parameters: a distance metric (epsilon) and a minimum number of points (MinPts) to define a cluster. DBSCAN works by grouping neighboring data points within the epsilon radius and satisfying the MinPts condition.

In summary, clustering algorithms provide a powerful tool for exploring and understanding the structure of your data. By applying these techniques to your datasets, you can uncover hidden patterns, relationships, and groupings that may not be immediately apparent through traditional analysis methods. This newfound knowledge can help you make more informed decisions and derive valuable insights from your data.

Dimensionality Reduction Techniques

In this section, we will explore dimensionality reduction techniques, a set of unsupervised learning methods that aim to reduce the number of features in your dataset while preserving its essential information. High-dimensional data can be challenging to work with due to the curse of dimensionality, which can lead to increased computational complexity, overfitting, and reduced model performance. By reducing the dimensionality of your data, you can improve the efficiency and effectiveness of your machine learning models.

We will discuss two popular dimensionality reduction techniques: Principal Component Analysis (PCA) and t-Distributed Stochastic Neighbor Embedding (t-SNE).

Principal Component Analysis (PCA)

PCA is a widely-used linear dimensionality reduction technique that aims to find a lower-dimensional representation of your data while preserving as much variance as possible. The main idea behind PCA is to project your data onto a new set of orthogonal axes (principal components) that capture the maximum amount of variance.

The principal components are linear combinations of the original features, with the first principal component accounting for the highest variance, the second principal component accounting for the second-highest variance, and so on. By selecting a subset of the principal components, you can reduce the dimensionality of your data while retaining most of its information.

To apply PCA to your data, follow these steps:

1. Standardize your dataset to have zero mean and unit variance for each feature.

2. Calculate the covariance matrix of the standardized data.
3. Compute the eigenvectors and eigenvalues of the covariance matrix.
4. Select the top k eigenvectors corresponding to the k largest eigenvalues, where k is the desired number of dimensions.
5. Project your data onto the subspace spanned by the selected eigenvectors.

For example, suppose you are working with a dataset containing the prices, sizes, and locations of real estate properties. By applying PCA, you can identify the main factors (principal components) that explain most of the variance in your dataset and use this reduced set of features to build more efficient and accurate machine learning models.

t-Distributed Stochastic Neighbor Embedding (t-SNE)

t-SNE is a non-linear dimensionality reduction technique that is particularly effective for visualizing high-dimensional data in two or three dimensions. The main goal of t-SNE is to preserve the local structure of your data, meaning that similar data points should remain close together in the lower-dimensional representation.

t-SNE works by minimizing the divergence between two probability distributions: one representing pairwise similarities in the high-dimensional space, and the other representing pairwise similarities in the low-dimensional space. The algorithm employs a gradient descent optimization method to find the optimal low-dimensional representation of your data.

Applying t-SNE to your data can help you visualize complex datasets, identify clusters or groupings, and gain insights into the underlying structure of your data. For instance, if you have a dataset containing images of handwritten digits, t-SNE can help you visualize the data in two dimensions, revealing distinct clusters corresponding to each digit.

In conclusion, dimensionality reduction techniques offer valuable tools for simplifying your data, improving model performance, and gaining insights into the structure and relationships within your dataset. By employing these methods, you can overcome the challenges posed by high-dimensional data and extract valuable information to inform your decision-making and analysis.

Anomaly Detection

In this section, we will discuss anomaly detection, an unsupervised learning technique used to identify unusual patterns or outliers in data. These outliers often represent rare events or observations that deviate significantly from the norm, which can provide valuable insights for various applications, such as fraud detection, network security, and quality control. We will explore three popular anomaly detection methods: Isolation Forest, Local Outlier Factor (LOF), and One-Class Support Vector Machines (OCSVM).

Isolation Forest

Isolation Forest is a tree-based anomaly detection algorithm that identifies outliers by isolating data points in the feature space. The main idea behind Isolation Forest is that anomalies are easier to separate from the rest of the data since they have fewer points in their vicinity. The algorithm works by recursively partitioning the data using randomly selected features and split values, creating a forest of isolation trees. The path length from the root to a data point in the trees is used to determine the anomaly score, with shorter path lengths indicating potential anomalies.

For example, you could apply Isolation Forest to detect fraudulent transactions in a financial dataset, helping you identify suspicious activities and take appropriate actions to prevent losses.

Local Outlier Factor (LOF)

LOF is a density-based anomaly detection method that measures the local deviation of a data point's density compared to its neighbors. The algorithm calculates the local density of each data point using a user-defined distance metric and the k-nearest neighbors approach. The LOF score for a data point is the ratio of the average local density of its k-nearest neighbors to its own local density. A higher LOF score indicates a potential outlier.

Suppose you're analyzing sensor data from an industrial production line to ensure consistent product quality. Using LOF, you can identify unusual sensor readings that may indicate equipment malfunctions or deviations in the manufacturing process.

One-Class Support Vector Machines (OCSVM)

OCSVM is an extension of the traditional Support Vector Machines (SVM) algorithm designed specifically for unsupervised anomaly detection. The main goal of OCSVM is to find a decision boundary that separates the majority of data points (normal observations) from the origin in the feature space while maximizing the margin. Data points on the other side of the decision boundary are considered anomalies.

For instance, you could apply OCSVM to detect intrusions in a computer network by analyzing network traffic patterns. Unusual traffic patterns that fall outside the decision boundary may indicate cyberattacks or unauthorized access attempts.

In conclusion, anomaly detection techniques offer valuable tools for identifying rare events or unusual patterns in your data. By applying these methods, you can uncover hidden insights, detect potential issues, and improve decision-making in various domains, such as finance, network security, and quality control. Understanding and utilizing these techniques will help you extract meaningful information from your data and tackle complex real-world challenges.

Association Rule Learning

In this section, we will explore Association Rule Learning, an unsupervised learning technique used to discover interesting relationships or associations between variables in large datasets. Association rules are widely used in various domains such as market basket analysis, recommendation systems, and web usage mining. We will focus on three popular Association Rule Learning algorithms: Apriori, Eclat, and FP-Growth.

Apriori Algorithm

The Apriori algorithm is a classic Association Rule Learning method that finds frequent itemsets in a dataset and generates association rules from these itemsets. Frequent itemsets are groups of items that appear together in a minimum number of transactions, as determined by a user-defined support threshold. The Apriori algorithm operates in a bottom-up approach, starting with single-item itemsets and extending them to larger itemsets while eliminating infrequent itemsets based on the support threshold.

For example, if you are analyzing customer purchase data from a supermarket, you can use the Apriori algorithm to discover frequent itemsets such as {bread, butter} or {milk, cereal}. Based on these frequent itemsets, you can then generate association rules like bread => butter, indicating that customers who buy bread are likely to buy butter as well.

Eclat Algorithm

The Eclat algorithm is a depth-first search-based Association Rule Learning method that focuses on the vertical representation of transactions. Instead of counting the occurrences of itemsets in the dataset like the Apriori algorithm, Eclat stores the Transaction Identifiers (TID) for each item and computes the intersections of TID sets to find frequent itemsets. The Eclat algorithm is generally faster than the Apriori algorithm, particularly for large datasets.

For instance, if you are analyzing movie viewership data from a streaming platform, the Eclat algorithm can help you find frequent itemsets like {action, sci-fi} or {romance, drama}, allowing you to develop personalized movie recommendations based on users' viewing patterns.

FP-Growth Algorithm

The FP-Growth (Frequent Pattern Growth) algorithm is an efficient Association Rule Learning method that utilizes a compact data structure called the FP-tree to represent the dataset. Unlike Apriori and Eclat, which rely on candidate generation and pruning, the FP-Growth algorithm eliminates the need for repeated database scans and candidate itemset generation. The algorithm constructs the FP-tree and mines frequent itemsets directly from the tree structure.

For example, if you are analyzing browsing data from an e-commerce website, you can use the FP-Growth algorithm to discover frequent itemsets like {electronics, accessories} or {books, stationery}. Based on these frequent itemsets, you can then create targeted promotions or cross-selling strategies to boost sales and customer satisfaction.

Association Rule Learning techniques offer insights into variable relationships and associations in data. They reveal hidden patterns, provide personalized recommendations, and aid data-driven decision-making in domains like retail, entertainment, and web analytics. Understanding these techniques uncovers valuable information and tackles real-world problems efficiently.

Autoencoders

In this section, we will delve into autoencoders, a type of artificial neural network used for unsupervised learning tasks, including dimensionality reduction and feature extraction. Autoencoders consist of two main components: an encoder that compresses the input data and a decoder that reconstructs the data from the compressed representation. We will discuss the structure, working principle, and practical applications of autoencoders.

Structure and Working Principle

An autoencoder consists of an input layer, one or more hidden layers, and an output layer. The encoder and decoder are connected through a bottleneck layer, also known as the latent space or code layer, which contains the compressed representation of the input data. The primary goal of an autoencoder is to minimize the reconstruction error, which is the difference between the original input and the reconstructed output. This is achieved by adjusting the weights of the neural network through a process called backpropagation and using an optimization algorithm, such as gradient descent.

During the training process, the autoencoder learns to extract the most important features from the input data and represent them in a lower-dimensional space, enabling efficient data compression and noise reduction. The decoder then uses this compressed representation to reconstruct the input data as accurately as possible.

Practical Applications

Autoencoders have a wide range of practical applications across various domains. Some examples include:

- **Dimensionality Reduction:** Autoencoders can be used to reduce the dimensionality of high-dimensional data, making it easier to visualize and analyze. For example, you can use an autoencoder to compress a large dataset of images or text data, retaining the most important features while reducing the computational complexity and storage requirements.

- **Feature Extraction:** Autoencoders can learn to extract meaningful features from raw data, which can then be used as input for other machine learning models. For instance, you can train an autoencoder on a dataset of unlabeled images to learn representative features, and then use these features to train a supervised classification model for image recognition tasks.

- **Denoising:** Autoencoders can be used to remove noise from data, improving the quality of the input for further analysis or processing. By training an autoencoder on noisy data and using the clean, reconstructed output, you can enhance the performance of subsequent machine learning models or data analysis techniques.

- **Anomaly Detection:** Autoencoders can be employed for anomaly detection tasks, as they learn to reconstruct normal data accurately, while struggling to reconstruct anomalous data points. By measuring the reconstruction error for each data point, you can identify potential outliers or anomalies in your dataset.

Autoencoders offer powerful unsupervised learning for tasks like dimensionality reduction, feature extraction, denoising, and anomaly detection. Mastering them uncovers hidden patterns, extracts valuable information, and solves real-world challenges.

Topic Modeling

In this section, we will explore topic modeling, an unsupervised learning technique used to discover hidden thematic structures in large collections of text documents. Topic modeling algorithms analyze the words in each document and group them into topics, enabling you to uncover the underlying themes and patterns in your data. We will focus on two popular topic modeling methods: Latent Dirichlet Allocation (LDA) and Non-negative Matrix Factorization (NMF).

Latent Dirichlet Allocation (LDA)

LDA is a generative probabilistic model that assumes each document in a corpus is a mixture of various topics and that each topic is a distribution over words. The primary goal of LDA is to estimate the topic-word distributions and document-topic distributions in the data. LDA uses a three-level hierarchical Bayesian model and relies on a process called Gibbs sampling to approximate the posterior distributions.

For example, if you have a large collection of news articles, you can use LDA to identify the main topics covered in the articles, such as politics, sports, or technology. This can help you organize and categorize the articles more effectively, enabling easier navigation and analysis.

Non-negative Matrix Factorization (NMF)

NMF is an algebraic method that decomposes a non-negative data matrix into two lower-dimensional non-negative matrices. In the context of topic modeling, NMF approximates the term-document matrix by the product of two lower-dimensional matrices, one representing the topics and the other representing the topic distribution for each document. NMF is an iterative algorithm that minimizes the reconstruction error between the original data matrix and the product of the two lower-dimensional matrices.

For instance, if you are analyzing a large collection of customer reviews, you can apply NMF to identify the key themes discussed by the customers, such as product quality, customer service, or pricing. This information can help you better understand customer sentiment and inform your business strategy.

Practical Applications

Topic modeling techniques have various practical applications across diverse domains. Some examples include:

- **Text Classification:** Topic modeling can be used to automatically classify documents into different categories based on their content. By extracting topics from a collection of documents, you can create a more structured representation of your data, making it easier to search, browse, and analyze.

- **Sentiment Analysis:** Topic modeling can be combined with sentiment analysis to understand the opinions and emotions expressed in text data. For example, you can analyze social media posts to identify popular topics of discussion and the sentiment associated with each topic, helping you gain insights into public opinion and trends.

- **Recommendation Systems:** Topic modeling can be employed to build content-based recommendation systems by identifying similar documents based on their topic distributions. For example, you can recommend articles or blog posts to users based on their reading history and the similarity of topics in the content.

- **Trend Analysis:** Topic modeling can be used to analyze the evolution of topics over time, allowing you to identify emerging trends and track changes in public discourse. For example, you can analyze news articles or academic publications to study the development of research fields or the shifting focus of media coverage.

In summary, topic modeling techniques offer valuable tools for discovering hidden structures and themes in large text datasets. By applying these methods, you can uncover meaningful insights, improve the organization and analysis of your data, and tackle complex real-world challenges in various domains, such as text classification, sentiment analysis, and trend analysis. Understanding and utilizing these techniques will help you extract valuable information from your data and enhance your ability to make data-driven decisions.

Module 6 Exam Questions

1. What is the primary goal of unsupervised machine learning?

a) Predicting a target variable based on input features

b) Discovering hidden patterns and structures in data

c) Classifying data into predefined categories

d) Training a model with labeled data

2. Which of the following is a clustering algorithm?

a) K-means

b) Linear regression

c) Decision tree

d) Support vector machine

3. Principal Component Analysis (PCA) is used for:

a) Clustering

b) Classification

c) Dimensionality reduction

d) Regression

4. In Association Rule Learning, the Apriori algorithm aims to find:

a) Frequent itemsets

b) Regression coefficients

c) Classification boundaries

d) Optimal clustering centroids

5. Which algorithm uses a depth-first search approach for Association Rule Learning?

a) Apriori

b) Eclat

c) FP-Growth

d) K-means

6. Autoencoders are primarily used for:

a) Supervised classification

b) Regression

c) Dimensionality reduction and feature extraction

d) Clustering

7. What is the objective of the Latent Dirichlet Allocation (LDA) algorithm in topic modeling?

a) To estimate topic-word and document-topic distributions

b) To find the optimal number of clusters

c) To classify documents into predefined categories

d) To approximate the posterior distributions using Gibbs sampling

8. What is the primary application of Non-negative Matrix Factorization (NMF)?

a) Clustering

b) Dimensionality reduction

c) Classification

d) Regression

9. Which of the following is a practical application of topic modeling?

a) Text classification

b) Image recognition

c) Speech recognition

d) Time series forecasting

10. Anomaly detection is often used to:

a) Predict a target variable based on input features

b) Identify potential outliers or anomalies in a dataset

c) Classify data into predefined categories

d) Train a model with labeled data

11. What is the primary goal of the K-means algorithm?

a) To find the optimal number of clusters

b) To predict a target variable based on input features

c) To partition data points into k clusters, minimizing the within-cluster sum of squares

d) To classify data into predefined categories

12. In which application domain are Association Rule Learning algorithms commonly used?

a) Market basket analysis

b) Image recognition

c) Speech recognition

d) Time series forecasting

13. Which of the following is NOT a use case for autoencoders?

a) Denoising

b) Anomaly detection

c) Supervised classification

d) Dimensionality reduction

14. In the context of unsupervised machine learning, what is the main difference between clustering and dimensionality reduction techniques?

a) Clustering groups similar data points together, while dimensionality reduction reduces the number of features

b) Clustering reduces the number of features, while dimensionality reduction groups similar data points together

c) Clustering predicts a target variable based on input features, while dimensionality reduction classifies data into predefined categories

d) Clustering classifies data into predefined categories, while dimensionality reduction predicts a target variable based on input features

15. Which of the following unsupervised learning techniques can be employed for anomaly detection?

a) K-means

b) PCA

c) Autoencoders

d) LDA

16. Which of the following unsupervised learning techniques is most suitable for text data analysis?

a) Topic modeling

b) Autoencoders

c) K-means

d) PCA

17. What is the primary purpose of the Eclat algorithm?

a) To find frequent itemsets

b) To perform dimensionality reduction

c) To cluster similar data points

d) To classify documents into predefined categories

18. Which of the following is a dimensionality reduction technique?

a) K-means clustering

b) Support vector machines

c) Principal Component Analysis (PCA)

d) Latent Dirichlet Allocation (LDA)

19. Which technique can be used in combination with topic modeling for sentiment analysis?

a) Clustering

b) Association Rule Learning

c) Anomaly Detection

d) Text Classification

20. What is the primary goal of t-Distributed Stochastic Neighbor Embedding (t-SNE)?

a) To visualize high-dimensional data in a lower-dimensional space

b) To partition data points into k clusters

c) To predict a target variable based on input features

d) To identify potential outliers or anomalies in a dataset

Answers:

1. b
2. a
3. c
4. a
5. b
6. c
7. a
8. b
9. a
10. b
11. c
12. a
13. c
14. a
15. c
16. a
17. a
18. c
19. d
20. a

Module 7. Deep Learning and Reinforcement Learning

Introduction to Deep Learning and Reinforcement Learning

Welcome to Module 7 on Deep Learning and Reinforcement Learning. In this module, you will learn about the fascinating world of deep learning and reinforcement learning, two advanced machine learning techniques that have revolutionized the way we approach complex problems and unlock the full potential of artificial intelligence.

Deep learning, a subset of machine learning, employs artificial neural networks to model and solve complicated tasks. By utilizing multiple layers of neurons, deep learning models can automatically learn intricate patterns and representations from large amounts of data. This ability has led to remarkable advances in various domains, such as computer vision, natural language processing, speech recognition, and more.

For instance, deep learning has empowered self-driving cars to detect and recognize objects like pedestrians, other vehicles, and traffic signs in real-time. In the realm of natural language processing, deep learning models have demonstrated impressive capabilities in generating human-like text, enabling applications like chatbots and AI-powered personal assistants.

Reinforcement learning, on the other hand, is a type of machine learning where an agent learns to make decisions by interacting with its environment. The agent receives feedback in the form of rewards or penalties and adjusts its actions to maximize the cumulative reward. This learning paradigm has been highly successful in addressing complex decision-making problems, ranging from robotics to game playing.

A famous example of reinforcement learning is Google DeepMind's AlphaGo, which made headlines when it defeated the world champion Go player, Lee Sedol, in 2016. This achievement was considered a significant milestone in AI, as Go is an exceptionally complex game with more possible board configurations than there are atoms in the observable universe. Reinforcement learning has also been applied to various other fields, including robotics, finance, and healthcare, where it has shown great promise in optimizing decision-making processes.

Exploring deep learning and reinforcement learning will allow you to acquire a deeper understanding of the cutting-edge techniques that are shaping the future of artificial intelligence. These methods offer powerful tools to tackle increasingly complex challenges and enable innovative solutions across numerous industries. This module will provide you with the knowledge and practical examples needed to harness the full potential of deep learning and reinforcement learning, paving the way for exciting new possibilities in your data science journey.

Introduction to Neural Networks

In this section, we will introduce you to the fascinating world of neural networks, the foundation of deep learning. Neural networks are inspired by the human brain's structure and function, aiming to mimic its remarkable ability to learn, recognize patterns, and adapt to new information.

A neural network is composed of interconnected artificial neurons or nodes, organized into layers. The primary layers in a neural network are the input layer, hidden layers, and the output layer. The input layer receives the initial data, while the output layer provides the final prediction or classification. Between these two layers, one or more hidden layers process and transform the input data to learn complex patterns and representations.

Each artificial neuron, also known as a node or unit, receives input from multiple sources, processes this input, and sends the output to other neurons in the network. The inputs are weighted, meaning they have associated weights that reflect their relative importance. As you train the neural network, these weights are adjusted to minimize the error between the network's predictions and the true target values.

Let's dive into a practical example to illustrate how neural networks can be used to solve real-world problems. Imagine you are working on a project to predict house prices based on various features, such as the number of bedrooms, square footage, and location. You could build a neural network to model the relationship between these features and the house prices.

First, you would define the structure of your neural network, including the number of input nodes (equal to the number of features), output nodes (one, since you are predicting a single value), and hidden layers (chosen based on the complexity of the problem). Next, you would initialize the weights randomly.

To train the neural network, you would feed it a dataset of historical house prices and their corresponding features. The neural network would iteratively process the input data, adjust the weights to minimize the error, and improve its predictions. Once the training is complete, you can use the neural network to predict the prices of new houses based on their features.

Activation functions play a crucial role in neural networks, as they introduce non-linearity into the model, enabling it to learn complex relationships between inputs and outputs. Common activation functions include the sigmoid, hyperbolic tangent (tanh), and rectified linear unit (ReLU) functions.

In summary, neural networks are powerful machine learning models capable of learning intricate patterns and representations in data. By understanding the core concepts of neural networks, such as layers, artificial neurons, weights, and activation functions, you can harness their potential to tackle a wide range of challenging problems across various domains.

Convolutional Neural Networks

In this section, we will introduce you to Convolutional Neural Networks (CNNs), a powerful and popular type of neural network designed to handle grid-like data, such as images, speech signals, and time-series data. CNNs have achieved remarkable success in various computer vision tasks, including image classification, object detection, and image generation.

The key innovation in CNNs is the use of convolutional layers, which are designed to automatically and adaptively learn spatial hierarchies of features. Convolutional layers consist of multiple filters, also known as kernels, which slide across the input data and perform element-wise multiplication followed by summation. This operation allows the CNN to learn local patterns in the data, such as edges, textures, and shapes in images.

Let's explore a practical example to understand how CNNs can be applied to real-world problems. Suppose you are working on a project to classify images of handwritten digits from the famous MNIST dataset. You could build a CNN to learn and recognize the unique patterns present in each digit.

To design your CNN, you would first create a sequence of layers, including convolutional layers, activation functions (e.g., ReLU), and pooling layers. Convolutional layers with multiple filters will learn different features from the input image, such as edges and corners. Pooling layers, such as max pooling or average pooling, are responsible for downsampling the data, reducing the spatial dimensions and computational complexity while preserving important features.

After the convolutional and pooling layers, you would add one or more fully connected layers, also known as dense layers, to perform the final classification. The last dense layer should have as many output nodes as there are classes (in the MNIST example, ten nodes for digits 0-9). You would typically use the softmax activation function in the final layer, as it provides a probability distribution over the classes.

Once you have defined the structure of your CNN, you would train it using a labeled dataset of handwritten digits and their corresponding class labels. During training, the CNN learns to recognize distinct features in each digit and adjusts the weights of the filters in the convolutional layers to minimize the classification error. After training, you can use the CNN to classify new images of handwritten digits accurately.

In summary, Convolutional Neural Networks are a specialized type of neural network designed to handle grid-like data, excelling at tasks involving images, speech signals, and time-series data. By understanding the principles behind CNNs, such as convolutional layers, pooling layers, and activation functions, you can leverage their power to solve complex problems in computer vision and other domains.

Recurrent Neural Networks

In this section, we will introduce you to Recurrent Neural Networks (RNNs), a unique type of neural network designed to handle sequential data, such as time series, natural language, and audio signals. RNNs have achieved great success in various tasks involving sequence-to-sequence learning, including machine translation, sentiment analysis, and speech recognition.

The defining feature of RNNs is their ability to maintain an internal state or memory, which allows them to capture information from previous time steps in the sequence. This is achieved through recurrent connections that loop back to the network's own hidden layers, enabling the RNN to retain information about past inputs and use it to inform future predictions.

Let's consider a practical example to understand how RNNs can be applied to real-world problems. Imagine you are working on a project to predict the sentiment of movie reviews based on the text. You could build an RNN to analyze the text and identify the sentiment as positive, negative, or neutral.

To design your RNN, you would first preprocess the text data by tokenizing it into words or subwords and encoding these tokens as numerical values. Next, you would define the structure of your RNN, including the input layer, one or more recurrent hidden layers, and the output layer. Popular types of recurrent layers include simple RNNs, Long Short-Term Memory (LSTM) layers, and Gated Recurrent Unit (GRU) layers. LSTM and GRU layers are particularly effective at handling long-range dependencies and mitigating the vanishing gradient problem, which can hamper the learning process in simple RNNs.

The output layer in your RNN should have as many nodes as there are classes (in the sentiment analysis example, three nodes for positive, negative, and neutral). The softmax activation function is commonly used in the final layer to provide a probability distribution over the classes.

Once you have defined the structure of your RNN, you would train it using a labeled dataset of movie reviews and their corresponding sentiment labels. During training, the RNN learns to recognize patterns in the text indicative of the sentiment and adjusts its weights to minimize the classification error. After training, you can use the RNN to predict the sentiment of new movie reviews.

In summary, Recurrent Neural Networks are a powerful type of neural network designed to handle sequential data by maintaining an internal state or memory. By understanding the core concepts behind RNNs, such as recurrent connections, hidden layers, and different types of recurrent layers like LSTMs and GRUs, you can harness their potential to tackle a wide range of challenging problems involving sequence-to-sequence learning.

Reinforcement Learning

In this section, we will introduce you to Reinforcement Learning (RL), a distinct type of machine learning where an agent learns to make decisions by interacting with an environment. The agent receives feedback in the form of rewards or penalties and aims to maximize the cumulative reward by choosing the best actions. Reinforcement learning has been highly successful in solving complex decision-making problems, ranging from game playing to robotics and finance.

The key components in a reinforcement learning setup are the agent, environment, states, actions, and rewards. The agent is the decision-maker, the environment represents the context in which the agent operates, states describe the current situation, actions are the possible choices the agent can make, and rewards signal the immediate feedback received after taking an action.

Reinforcement learning algorithms can be classified into model-free and model-based methods. Model-free methods, such as Q-learning and policy gradients, learn directly from interactions with the environment without explicitly modeling the underlying dynamics. Model-based methods, on the other hand, attempt to learn a model of the environment's dynamics and use it to plan and make decisions.

Let's explore a practical example to understand how reinforcement learning can be applied to real-world problems. Imagine you are developing an intelligent agent to control a self-driving car. The goal is to teach the agent to navigate safely and efficiently from a starting point to a destination while avoiding obstacles and adhering to traffic rules.

In this scenario, the agent would be the self-driving car's control system, and the environment would include the road, traffic conditions, and other vehicles. The states could be represented by the car's position, speed, and nearby objects, while the actions might include accelerating, braking, turning, and changing lanes.

The rewards would be designed to encourage safe and efficient driving behavior, such as staying within speed limits, maintaining a safe distance from other vehicles, and reaching the destination quickly.

To train your agent, you could employ a reinforcement learning algorithm, such as Deep Q-Networks (DQN) or Proximal Policy Optimization (PPO), which combine traditional RL methods with deep learning techniques. The agent would learn by repeatedly interacting with a simulated environment, adjusting its policy based on the rewards received to improve its decision-making over time. Once the agent has been trained, you can deploy it in the self-driving car to navigate autonomously in real-world conditions.

In summary, Reinforcement Learning is a powerful approach to machine learning that focuses on decision-making through interaction with an environment. By understanding the core concepts behind reinforcement learning, such as agents, environments, states, actions, and rewards, you can harness its potential to solve complex problems in various domains, from autonomous vehicles to finance and healthcare.

Transfer Learning

In this section, we will delve deeper into transfer learning, an essential technique in deep learning that enables you to harness the knowledge gained from one task to improve performance on a new, related task. This approach can save significant time and computational resources, as well as lead to better model performance, especially when dealing with smaller datasets.

Transfer learning is particularly effective in scenarios where the new task shares similar features or patterns with the original task. It is commonly applied in computer vision and natural language processing, where neural network architectures can be reused with minor modifications.

Here are the key steps to leveraging transfer learning effectively:

- **Select a pre-trained model:** First, choose a suitable pre-trained model, such as ResNet or VGG for computer vision tasks, or BERT and GPT for natural language processing tasks. These models have already learned useful features on large-scale datasets and are available in popular deep learning libraries like TensorFlow and PyTorch.

- **Remove the final layer(s):** Since the pre-trained model was designed for a different task, you need to remove the final classification layer(s) to adapt the model to your specific problem.

- **Add new layers:** Next, add new layers to the model to match the output size of your target task. For instance, in a plant classification example, the new output layer should have as many nodes as there are plant classes.

- **Fine-tune the model:** Finally, train the model on your dataset. You have two primary options for this step:

    - **Feature extraction:** Update only the newly added layers while keeping the pre-trained layers' weights fixed. This approach is suitable when your dataset is small or has limited similarity to the original task.

- **Fine-tuning:** Update the entire network, including the pre-trained layers. This is appropriate when your dataset is larger or shares more similarities with the original task. In this case, you may want to use a smaller learning rate to preserve the learned features and prevent overfitting.

Let's consider a practical example in the context of computer vision. Suppose you are building a model to classify images of various dog breeds. Training a deep neural network from scratch can be computationally expensive and require a large labeled dataset. Instead, you could use a pre-trained network like ResNet, which has already learned useful features from a large-scale dataset like ImageNet.

Following the steps mentioned above will allow you to adapt the ResNet model to your dog breed classification task. Once fine-tuned, your model will be able to classify dog breeds accurately, leveraging the features learned from the original ImageNet dataset.

In the context of natural language processing, transfer learning can also be applied using models like BERT or GPT. For example, if you are building a sentiment analysis model, you can use a pre-trained BERT model, remove its final classification layer, and add a new layer for binary classification (positive or negative sentiment). Fine-tuning the model on your sentiment analysis dataset will enable it to learn the nuances of your specific task while leveraging the language understanding capabilities of the pre-trained BERT model.

In summary, transfer learning is a powerful technique that allows you to leverage the knowledge gained from one task to improve performance on a new, related task. By understanding how to select, adapt, and fine-tune pre-trained models, you can save significant time and computational resources while achieving better results, especially when working with smaller datasets. Transfer learning is widely applicable across domains, including computer vision, natural language processing, and beyond.

Generative Models

In this section, we will explore Generative Models in greater depth, diving into the capabilities of these powerful neural networks designed to generate new data samples that resemble the original data distribution. These models have achieved remarkable success in a wide range of applications, such as image synthesis, text generation, and audio creation.

There are two primary types of generative models: Variational Autoencoders (VAEs) and Generative Adversarial Networks (GANs). We will investigate both types, discussing their inner workings and providing practical examples to help you understand their potential applications.

- **Variational Autoencoders (VAEs):** VAEs are a type of autoencoder that learns to encode data into a lower-dimensional latent space and then decode it back into the original space. VAEs enforce a specific structure on the latent space, enabling them to generate new data samples by sampling from this space. VAEs have been used in various applications, such as generating images, interpolating between data points, and data compression.

  A practical example of VAEs involves generating new images of faces. Suppose you have a dataset of celebrity faces and want to create new, realistic images of faces that do not exist. You could train a VAE on the celebrity dataset, learning to encode and decode facial images. Once trained, you can sample from the latent space to generate new, diverse, and realistic images of faces that were not part of the original dataset.

- **Generative Adversarial Networks (GANs):** GANs consist of two neural networks, a generator and a discriminator, which compete against each other in a game-like setting. The generator learns to create realistic data samples, while the discriminator learns to distinguish between real samples from the dataset and fake samples produced by the generator. As the training progresses, both networks improve, leading to the generator creating increasingly realistic samples. GANs have been used to generate high-quality images, create artwork, and synthesize realistic human faces, among other applications.

A practical example of GANs is the generation of photorealistic images, such as high-resolution landscapes. Suppose you have a dataset of landscape images and want to create new, visually appealing images of nature scenes. You could train a GAN on the dataset, with the generator producing new images and the discriminator evaluating their quality. As the training progresses, the generator will learn to create increasingly realistic and diverse landscape images, suitable for various purposes like digital art or virtual environments.

Another exciting application of GANs is the generation of realistic human faces, as seen in projects like NVIDIA's StyleGAN. By training a GAN on a large dataset of human faces, you can create synthetic faces that are virtually indistinguishable from real ones. These generated faces can be used in various scenarios, such as creating virtual characters for video games or generating diverse avatars for online platforms.

In summary, Generative Models are neural networks designed to generate new data samples that resemble the original data distribution. By gaining a deeper understanding of the core concepts behind VAEs and GANs, you can harness their potential to tackle a wide range of applications, from image synthesis to text and audio generation. These models open up exciting possibilities for creativity, innovation, and problem-solving across various domains, offering you the tools to generate unique and realistic content for a plethora of use cases.

Attention Mechanisms

In this section, we will delve deeper into Attention Mechanisms, powerful techniques that enable neural networks to selectively focus on specific parts of the input, leading to improved performance and interpretability. We will discuss the core concepts behind attention mechanisms, the types of attention used in deep learning, and provide practical examples to demonstrate their impact on real-world applications, particularly in natural language processing tasks.

The central idea behind attention mechanisms is to compute a weighted sum of input elements, where the weights represent the importance or relevance of each element for the current output. Instead of processing the entire input sequence at once, the model can dynamically focus on the most relevant parts, which is particularly useful for handling long sequences and complex data structures.

There are several types of attention mechanisms, including:

- **Global Attention:** The model computes attention scores for all input elements at each time step, allowing it to attend to the entire input sequence. This is well-suited for tasks like machine translation, where long-range dependencies play a crucial role.

- **Local Attention:** The model computes attention scores for a small, fixed-size window of input elements at each time step. This is useful for tasks where local context is more important than global context, such as part-of-speech tagging or named entity recognition.

A widely-used attention mechanism is the "Scaled Dot-Product Attention," which is a key component of the Transformer architecture. The Transformer has become the basis for many state-of-the-art natural language processing models, such as BERT, GPT, and T5.

To illustrate the power of attention mechanisms, let's consider a practical example involving machine translation. Suppose you are building a model to translate English sentences into French. In a traditional sequence-to-sequence model without attention, the model would encode the entire input sentence into a fixed-size vector and then decode it into the target language. This approach can lead to difficulties in capturing long-range dependencies and handling longer input sequences.

By incorporating attention mechanisms, your model can selectively focus on specific words or phrases in the English sentence while generating the corresponding French words. This dynamic, context-aware processing enables the model to better capture the relationships between words and produce more accurate translations, especially for longer sentences or complex language structures.

Another practical example is abstractive text summarization. In this task, you want to create a concise summary of a given text while preserving its core meaning. By using attention mechanisms, your model can focus on the most important parts of the input text, enabling it to generate summaries that are both coherent and informative.

In summary, Attention Mechanisms enhance neural networks by selectively focusing on input parts, improving performance and interpretability. Grasping core concepts, types, and applications in sequence-to-sequence models enables designing advanced natural language processing systems, unlocking potential in machine translation, text summarization, and question-answering systems.

Reinforcement Learning Exploration-Exploitation Trade-off

In this section, we will discuss the exploration-exploitation trade-off, a central challenge in reinforcement learning that determines how an agent should balance between exploring new actions and exploiting actions that are already known to yield good rewards. Understanding this trade-off is crucial for designing effective reinforcement learning algorithms that can adapt and learn in complex environments.

The exploration-exploitation trade-off is critical because it directly impacts the agent's ability to learn optimal policies. Exploring new actions allows the agent to gather information about the environment, potentially discovering better strategies. On the other hand, exploiting known actions ensures the agent receives rewards for its current knowledge. Striking the right balance between exploration and exploitation is essential for achieving optimal performance.

There are several strategies to address the exploration-exploitation trade-off in reinforcement learning:

- **Epsilon-Greedy:** In this approach, you define a parameter epsilon (ε), representing the probability of taking a random action for exploration. With a probability of 1-ε, the agent chooses the action with the highest estimated reward, exploiting its current knowledge. Over time, you can gradually decrease ε to favor exploitation as the agent's knowledge of the environment improves.

  A practical example of the epsilon-greedy approach can be found in a simple game like tic-tac-toe. At the beginning of the learning process, your agent might make random moves (high ε) to explore the environment and understand the game's dynamics. As it gains experience and learns the optimal moves, you can decrease ε, allowing the agent to exploit its knowledge and make better decisions, ultimately playing optimally.

- **Upper Confidence Bound (UCB):** The UCB algorithm addresses the trade-off by assigning a confidence bound to each action, estimating the potential reward. Actions with higher uncertainty or less explored actions will have wider confidence bounds. The agent selects the action with the highest upper confidence bound, balancing exploration and exploitation based on uncertainty.

  A practical application of the UCB algorithm is in the multi-armed bandit problem, where you have several slot machines (bandits) with unknown reward probabilities. Your goal is to maximize your total reward by pulling the arms in an optimal sequence. The UCB algorithm helps you balance between trying different arms (exploration) and sticking to the arm with the highest estimated reward (exploitation).

- **Thompson Sampling:** This Bayesian approach models the uncertainty of the expected reward for each action using a probability distribution. At each time step, the agent samples from these distributions to select an action, updating the distributions as new rewards are observed. This method balances exploration and exploitation based on the agent's uncertainty about the true rewards.

Consider an online advertising scenario, where your goal is to display ads with the highest click-through rate. Thompson Sampling can help you balance between showing different ads (exploration) and focusing on the ads with the highest estimated click-through rates (exploitation) by maintaining and updating probability distributions for each ad's performance.

In summary, understanding and addressing the exploration-exploitation trade-off is critical for designing effective reinforcement learning agents that can learn and adapt in complex environments. By familiarizing yourself with various strategies like epsilon-greedy, UCB, and Thompson Sampling, you can make informed decisions on how to balance exploration and exploitation in your reinforcement learning projects, leading to better performance and more optimal policies in a wide range of applications.

Multi-Agent Reinforcement Learning

In this section, we will delve into Multi-Agent Reinforcement Learning (MARL), an extension of reinforcement learning that deals with multiple learning agents interacting with each other in a shared environment. Understanding MARL is crucial when developing AI systems that must cooperate, compete, or negotiate with other agents, whether they are other AI systems or humans.

Multi-Agent Reinforcement Learning poses unique challenges and opportunities compared to single-agent reinforcement learning:

- **Cooperation:** In cooperative settings, agents must work together to achieve a shared goal, often requiring them to communicate and coordinate their actions. Learning cooperative strategies can be challenging, as the success of an individual agent depends on the actions of its teammates.
- **Competition:** In competitive environments, agents aim to maximize their own rewards at the expense of other agents. The challenge in such settings is to learn strategies that outperform opponents while being robust to their adaptive behaviors.
- **Mixed Cooperation and Competition:** Many real-world scenarios involve a mix of cooperative and competitive elements, making it crucial for agents to learn how to balance these aspects effectively.

To address these challenges, various approaches have been developed for MARL:

- **Independent Q-Learning:** Each agent learns its own Q-values independently, treating other agents as part of the environment. While this approach is simple, it can struggle in complex scenarios, as the presence of other learning agents leads to a non-stationary environment.

- **Joint Action Learning:** Agents learn joint action values, considering the actions of all agents simultaneously. This approach can capture the interdependencies between agents, but it suffers from scalability issues due to the exponential growth of the action space with the number of agents.
- **Centralized Training with Decentralized Execution:** During training, a central controller has access to the full state and actions of all agents. However, during execution, each agent makes decisions based on its local observations. This approach leverages global information during training while maintaining scalability during execution.

Now, let's look at some practical examples of MARL applications:

- **Autonomous Vehicles:** In a traffic scenario, multiple self-driving cars must cooperate to ensure smooth traffic flow and avoid collisions. They may also compete for resources such as parking spots or optimal lanes. MARL techniques can be employed to teach self-driving cars to effectively navigate complex traffic situations while interacting with other vehicles.
- **Robotic Soccer:** In this cooperative-competitive domain, two teams of robots play soccer against each other. Each team's agents must cooperate to pass the ball, defend, and score goals, while also competing against the opposing team. Multi-agent reinforcement learning can be used to develop sophisticated strategies and improve team coordination.
- **Smart Grid Management:** In a smart grid, multiple agents, such as electricity producers and consumers, interact to balance energy supply and demand. Producers must

cooperate to ensure a stable power supply, while consumers may compete for the best prices. MARL can help optimize energy distribution and pricing in such complex, multi-agent environments.

- **Financial Markets:** In an automated trading system, multiple trading agents compete to maximize profits while adapting to the actions of other market participants. MARL can be used to develop adaptive trading strategies that consider the dynamic and competitive nature of financial markets.

In conclusion, Multi-Agent Reinforcement Learning is a crucial area to understand when developing AI systems that interact with other agents in complex environments. By studying various MARL approaches and their applications in cooperative, competitive, and mixed settings, you can create more advanced and effective AI systems for a wide range of real-world problems.

Module 7 Exam Questions

1. What is the primary goal of reinforcement learning?

a) Supervised learning

b) Maximizing cumulative reward

c) Minimizing classification error

d) Clustering data points

2. Which of the following is NOT a type of neural network covered in this module?

a) Convolutional Neural Network

b) Recurrent Neural Network

c) Support Vector Machine

d) Feedforward Neural Network

3. What is the primary focus of the epsilon-greedy strategy in reinforcement learning?

a) Balancing exploration and exploitation

b) Maximizing exploration only

c) Maximizing exploitation only

d) Reducing the state space

4. Which Python library is commonly used for data manipulation and analysis in data science?

a) NumPy

b) Pandas

c) Matplotlib

d) Seaborn

5. What does the term "transfer learning" refer to in deep learning?

a) Combining multiple models

b) Applying knowledge learned from one task to another

c) Transferring data between different sources

d) Sharing weights between layers in a neural network

6. What is the purpose of attention mechanisms in deep learning models?

a) Optimizing model parameters

b) Focusing on relevant parts of the input

c) Reducing overfitting

d) Speeding up training

7. In the context of reinforcement learning, what is the exploration-exploitation trade-off?

a) Balancing the learning rate and discount factor

b) Balancing the rewards and penalties

c) Balancing the time spent learning and executing actions

d) Balancing the choice between trying new actions and using known actions

8. What type of machine learning algorithm is primarily used for dimensionality reduction?

a) Regression

b) Classification

c) Clustering

d) None of the above

9. What is the primary difference between supervised and unsupervised machine learning?

a) Supervised learning uses labeled data, while unsupervised learning uses unlabeled data

b) Supervised learning is used for classification, while unsupervised learning is used for regression

c) Supervised learning is deterministic, while unsupervised learning is probabilistic

d) Supervised learning is rule-based, while unsupervised learning is instance-based

10. Which of the following is NOT a key skill for data scientists?

a) Programming

b) Statistics

c) Graphic design

d) Machine learning

11. In reinforcement learning, which term represents the immediate reward an agent receives after taking an action?

a) Value function

b) Q-function

c) State

d) Reward

12. Which deep learning model is especially suitable for handling sequential data?

a) Convolutional Neural Networks

b) Feedforward Neural Networks

c) Recurrent Neural Networks

d) Boltzmann Machines

13. What is the primary goal of data visualization?

a) Summarizing data

b) Cleaning data

c) Communicating insights effectively

d) Transforming data

14. What is the primary purpose of hypothesis testing in statistics?

a) Describing data

b) Predicting future data points

c) Making inferences about a population

d) Calculating probabilities

15. Which of the following is a common application of reinforcement learning?

a) Image classification

b) Game playing

c) Sentiment analysis

d) Time series forecasting

16. In multi-agent reinforcement learning, which of the following is a unique challenge?

a) Cooperation

b) Competition

c) Mixed Cooperation and Competition

d) All of the above

17. Which of the following is a dimensionality reduction technique?

a) Linear regression

b) K-means clustering

c) Principal Component Analysis (PCA)

d) Decision Trees

18. In a convolutional neural network, which layer is responsible for detecting local patterns in the input data?

a) Fully connected layer

b) Convolutional layer

c) Pooling layer

d) Input layer

19. What is the primary advantage of using transfer learning in deep learning?

a) Reduced computational cost

b) Improved model accuracy

c) Faster training times

d) All of the above

20. What is the primary objective of data cleaning in the data science process?

a) Visualizing data

b) Identifying and correcting errors and inconsistencies in the data

c) Transforming data into a suitable format

d) Generating insights from the data

Answers:

1. b
2. c
3. a
4. b
5. b
6. b
7. d
8. d
9. a
10. c
11. d
12. c
13. c
14. c
15. b
16. d
17. c
18. b
19. d
20. b

Module 8. Data Wrangling and Preprocessing

Introduction to Data Wrangling and Preprocessing

Welcome to Module 8 on Data Wrangling and Preprocessing. In this module, you will learn the essential techniques and processes used to clean, transform, and prepare data for analysis and modeling in data science projects. Data wrangling and preprocessing are critical steps in the data science pipeline, as the quality of your data has a significant impact on the accuracy and reliability of your models and insights.

Data in the real world is often messy, incomplete, and inconsistent. It can come from multiple sources and formats, making it challenging to work with. This module will teach you how to tackle these common data-related issues and ensure that your data is in the best possible shape for further analysis and modeling.

Practical examples will be provided throughout the module to demonstrate the importance of data wrangling and preprocessing. You will be exposed to various real-world scenarios where data preprocessing techniques are applied to improve the quality of data and, subsequently, the performance of machine learning algorithms.

For instance, you might encounter a dataset containing information about customer purchases, but the data may be riddled with missing values, inconsistencies, and duplicate entries. By learning the data wrangling techniques presented in this module, you will be able to clean and preprocess the dataset effectively, ensuring that your analysis is based on accurate and reliable information.

As you progress through this module, you will gain hands-on experience with various data manipulation and preprocessing techniques, enhancing your understanding of the vital role they play in data science. By the end of this module, you will have the necessary skills to confidently approach any dataset and prepare it for further analysis or modeling, ensuring the success of your data science projects.

Data Collection and Importing

The first step in any data science project is data collection and importing. This involves gathering data from various sources and loading it into your programming environment, such as Python, for further processing and analysis. In this section, we will explore different data sources, techniques to collect data, and methods to import the data into your workspace.

Data Sources:

Data can come from a wide range of sources, including:

- **Files:** You may collect data from files such as CSV, Excel, JSON, or XML. These files can be stored locally on your computer or remotely on a server.
- **Databases:** Data can be fetched from relational databases (e.g., MySQL, PostgreSQL) or non-relational databases (e.g., MongoDB, Cassandra).
- **APIs:** Many organizations provide access to their data through Application Programming Interfaces (APIs). By interacting with these APIs, you can request and receive data in a structured format.
- **Web scraping:** Sometimes, data is not readily available through APIs or files, and you may need to extract information from websites using web scraping techniques.

Data Collection Techniques

Depending on the data source, you may employ different data collection techniques:

- **Reading Files:** To collect data from files, you will need to read and parse the contents of the file. Depending on the file format, you may use built-in Python functions or specialized libraries.

- **Querying Databases:** When collecting data from databases, you will typically write SQL (Structured Query Language) queries to fetch specific data from tables. For non-relational databases, you may use the database's query language or API.

- **Interacting with APIs:** To access data through APIs, you will need to send HTTP requests and parse the response, which is often in JSON or XML format. Many APIs require authentication, so you may need to obtain an API key or token.

- **Web Scraping:** Web scraping involves extracting data from websites using a combination of HTTP requests and parsing HTML content. You will often use libraries like Beautiful Soup or Scrapy to help extract the data of interest.

Data Importing

Once you have collected your data, you will need to import it into your programming environment. In Python, you can use various libraries to import data, depending on the data format and source:

- For CSV and Excel files, you can use the pandas library to read the data into a DataFrame, which is a powerful data structure for manipulating tabular data.

- For JSON and XML data, you can use the built-in json and xml libraries, respectively, to parse the contents and convert them into native Python data structures like dictionaries and lists.

- When working with databases, you may use libraries like SQLAlchemy or PyMySQL for relational databases and PyMongo or Cassandra-driver for non-relational databases.

- If you collect data through web scraping or APIs, you may need to preprocess the data to convert it into a suitable format (e.g., DataFrame) for further analysis.

In summary, data collection and importing is a crucial step in the data science process, and by mastering these techniques, you can ensure that your data is properly gathered and prepared for the subsequent stages of your project.

Data Cleaning

Data cleaning is an essential step in the data wrangling process, as it ensures that your dataset is accurate, consistent, and free of errors before analysis and modeling. In this section, we will discuss common data quality issues, techniques to identify and address them, and practical examples to illustrate the importance of data cleaning.

Common Data Quality Issues

Your dataset may have various quality issues, such as:

- **Missing Values:** Data points might be missing or incomplete for various reasons, such as data entry errors, equipment malfunctions, or respondents skipping questions in surveys.

- **Duplicate Records:** Duplicate entries may occur due to data entry errors, merging datasets, or other reasons, leading to incorrect analysis and conclusions.

- **Inconsistent Formats:** Data can be represented in different formats, units, or encodings, making it challenging to analyze and compare the data consistently.

- **Outliers:** Outliers are data points that are significantly different from the rest of the dataset. They can be due to measurement errors, data entry mistakes, or genuine deviations.

- **Incorrect Values:** Data points may contain incorrect values due to data entry errors, measurement errors, or other issues.

Data Cleaning Techniques

To address these data quality issues, you can use the following techniques:

- **Handling Missing Values:** Depending on the nature of the missing data and your analysis goals, you may choose to fill in missing values with a default value, the mean, median, or mode of the column, or use more advanced techniques like interpolation or regression imputation. Alternatively, you may decide to remove rows or columns with missing data, but this should be done cautiously to avoid losing valuable information.

- **Removing Duplicates:** To eliminate duplicate records, you can use methods such as identifying unique identifiers, comparing record similarity, or using built-in functions in data manipulation libraries like pandas.

- **Standardizing Formats:** You should convert your data into a consistent format, unit, or encoding to facilitate analysis. This may involve parsing dates and times, converting units of measurement, or mapping categorical values to a standard representation.

- **Identifying and Handling Outliers:** Outliers can be detected using visualizations (e.g., box plots) or statistical methods (e.g., Z-score, IQR). Once identified, you may decide to correct, remove, or keep the outliers, depending on the cause and context of the outlier.

- **Validating and Correcting Values:** To ensure that your data is accurate, you can use techniques like data validation rules, cross-referencing with other sources, or manual inspection. You may need to correct or remove incorrect values based on your findings.

Practical Example

Imagine you have a dataset containing information about customers and their purchases. Upon inspection, you find that some customer names have inconsistent capitalization, there are duplicate records for some customers, and certain purchase dates are missing or in different formats.

To clean this dataset, you would:

1. Standardize the capitalization of customer names.
2. Identify and remove duplicate customer records.
3. Fill in missing purchase dates or remove records with missing dates, depending on the context and impact on your analysis.
4. Convert all purchase dates to a consistent format, such as YYYY-MM-DD.

Performing these data cleaning steps will allow you to ensure that your dataset is accurate, consistent, and ready for further analysis and modeling.

Data Transformation and Feature Engineering

Data transformation and feature engineering are vital steps in the data preprocessing process, as they enable you to create meaningful and useful features from raw data, which can significantly improve the performance of your models. In this section, we will discuss various data transformation techniques, the concept of feature engineering, and practical examples to demonstrate their importance.

Data Transformation Techniques

Data transformation involves modifying your data to improve its quality, reduce complexity, or create new features. Some common data transformation techniques include:

- **Scaling:** Scaling adjusts the range of your numerical features to prevent features with large values from dominating the analysis or model. Techniques include min-max scaling, standard scaling (z-score), and robust scaling.

- **Encoding:** Encoding is the process of converting categorical data into numerical values that can be used by machine learning algorithms. Common encoding methods include one-hot encoding, ordinal encoding, and binary encoding.

- **Discretization:** Discretization involves converting continuous numerical features into discrete categories or bins. This can help reduce noise and improve model performance, especially for linear models.

- **Log and Power Transforms:** Logarithmic and power transformations can help stabilize variance, reduce skewness, and improve the distribution of your data, making it more suitable for analysis and modeling.

Feature Engineering

Feature engineering is the process of creating new features or modifying existing ones to enhance the quality and predictive power of your dataset. Some common feature engineering techniques include:

- **Domain-Specific Transformations:** Based on your domain knowledge, you may create new features by combining or transforming existing features in a meaningful way. For example, in a retail dataset, you could calculate the average purchase amount per customer.

- **Interaction Features:** Interaction features capture the combined effect of two or more features on the target variable. For example, you could create an interaction feature between the age and income of customers to better predict their purchasing behavior.

- **Polynomial Features:** Polynomial features are derived by taking powers or combinations of existing features. These can help capture non-linear relationships between features and the target variable.

- **Feature Selection:** Feature selection involves identifying the most relevant features for your analysis or model, which can help reduce overfitting, improve model interpretability, and reduce computational complexity.

Practical Example

Imagine you have a dataset containing information about houses, including their size, age, number of rooms, and sale price. You suspect that the price per square foot might be a useful feature for predicting the sale price.

To create this new feature, you would:

1. Divide the sale price by the size of the house to calculate the price per square foot.

2. Add this new feature to your dataset.

Additionally, you might decide to apply scaling to the size, age, and number of rooms features to ensure they have a similar range of values, making them more suitable for analysis and modeling.

Performing these data transformation and feature engineering steps will allow you to create meaningful features that capture the underlying patterns and relationships in your data, improving the performance of your analysis and models.

Handling Imbalanced Data

Imbalanced data is a common issue in classification problems, where the distribution of classes in the target variable is uneven. This can lead to biased model performance, as the model may be more likely to predict the majority class. In this section, we will discuss techniques to handle imbalanced data and practical examples to demonstrate their importance.

Techniques for Handling Imbalanced Data

- **Resampling:** You can balance the class distribution by either oversampling the minority class, undersampling the majority class, or a combination of both. Oversampling involves creating copies of instances from the minority class, while undersampling involves removing instances from the majority class. These methods should be used carefully to avoid overfitting or loss of information.

- **Synthetic Data Generation:** Methods like Synthetic Minority Over-sampling Technique (SMOTE) or Adaptive Synthetic (ADASYN) generate synthetic instances of the minority class by interpolating between existing instances. This can help increase the representation of the minority class without overfitting.

- **Using Different Evaluation Metrics:** Imbalanced datasets can lead to misleading accuracy scores. Instead, consider using evaluation metrics that are more robust to class imbalance, such as precision, recall, F1-score, or the area under the Receiver Operating Characteristic (ROC) curve.

- **Cost-Sensitive Learning:** You can assign different misclassification costs to the majority and minority classes, making the model more sensitive to the minority class. This can be done by adjusting class weights or modifying the learning algorithm itself.

- **Ensemble Methods:** Ensemble methods, such as bagging and boosting, can be adapted to handle imbalanced data. For example, you can use Random UnderSampling (RUS) with bagging or SMOTE with boosting to improve the performance on minority class instances.

Practical Example

Imagine you have a dataset for predicting fraudulent transactions, where the majority of instances are non-fraudulent transactions, and only a small percentage are fraudulent. Due to the class imbalance, a model trained on this dataset may have high accuracy, but fail to identify the majority of fraudulent transactions.

To address this issue, you could:

1. Apply SMOTE to generate synthetic fraudulent instances, balancing the class distribution.
2. Train a model using cost-sensitive learning, assigning a higher misclassification cost to fraudulent transactions.
3. Evaluate the model using the F1-score or the area under the ROC curve to obtain a more accurate representation of its performance on both classes.

Applying these techniques will allow you to improve the performance of your model on the minority class, ensuring that it is more effective at identifying fraudulent transactions in the presence of imbalanced data.

Text Data Preprocessing

Text data preprocessing is a crucial step when working with natural language data, as it prepares the raw text for analysis and modeling. Proper preprocessing can significantly improve the performance of your text-based machine learning models. In this section, we will discuss various text preprocessing techniques and provide practical examples to demonstrate their importance.

Text Preprocessing Techniques

- **Tokenization:** Tokenization is the process of breaking down text into individual words, phrases, or tokens. This allows you to analyze and model text at different levels, such as word-level or sentence-level. Common tokenization methods include whitespace-based tokenization, regular expression-based tokenization, and more advanced techniques like using pre-trained tokenizers from natural language processing libraries.

- **Stopword Removal:** Stopwords are common words that carry little meaning and can add noise to your analysis. Removing stopwords can help reduce the dimensionality of your data and improve the performance of your model. Common stopwords include "a", "an", "the", "and", "in", and so on.

- **Stemming and Lemmatization:** Stemming and lemmatization are techniques used to reduce words to their base or root form. Stemming is a more aggressive approach that removes word affixes, while lemmatization considers the context and part of speech to derive the base form. Both techniques can help improve the consistency of your text data and reduce dimensionality.

- **Text Cleaning:** Cleaning text involves removing irrelevant or noisy elements such as special characters, numbers, or

HTML tags. This helps simplify your text data and makes it more suitable for analysis and modeling.

- **Vectorization:** Vectorization is the process of converting text data into numerical representations that can be used by machine learning algorithms. Common vectorization methods include Bag of Words (BoW), Term Frequency-Inverse Document Frequency (TF-IDF), and word embeddings (e.g., Word2Vec, GloVe, or BERT).

Practical Example

Imagine you have a dataset containing customer reviews for a product, and you want to build a sentiment analysis model to classify the reviews as positive or negative. To preprocess the text data, you would:

1. Tokenize the reviews into words, using a tokenizer from a natural language processing library, such as NLTK or spaCy.
2. Remove stopwords from the tokenized words to reduce noise and dimensionality.
3. Apply lemmatization to the remaining words to ensure consistency and further reduce dimensionality.
4. Clean the text by removing special characters, numbers, and any irrelevant elements.
5. Convert the cleaned and preprocessed text into numerical vectors using a vectorization method, such as TF-IDF or word embeddings.

With the preprocessed text data, you can now train and evaluate sentiment analysis models, such as logistic regression, support vector machines, or deep learning models like recurrent neural networks (RNNs) or transformers. By performing these text preprocessing steps, you can improve the performance and effectiveness of your text-based machine learning models, ensuring they capture meaningful patterns and relationships within the natural language data.

Time Series Data Preprocessing

Time series data preprocessing is essential when working with data that is collected or recorded over time, such as stock prices, weather data, or sensor readings. Proper preprocessing can help ensure that your time series models effectively capture the underlying temporal patterns and relationships. In this section, we will discuss various time series preprocessing techniques and provide practical examples to demonstrate their importance.

Time Series Preprocessing Techniques

- **Resampling:** Resampling involves changing the frequency of your time series data, which can be helpful in addressing issues related to irregular sampling or handling data with multiple time scales. Common resampling methods include upsampling (increasing the frequency) and downsampling (decreasing the frequency). For example, you can resample daily stock prices to weekly or monthly frequencies, depending on your analysis needs.

- **Handling Missing Values:** Time series data often contains missing values, which can result from data collection issues or irregular sampling. You can handle missing values using various techniques, such as forward or backward filling, linear interpolation, or more advanced methods like seasonal decomposition or state space models. Choose the method that best preserves the underlying temporal structure of your data.

- **Removing Seasonality and Trend:** Time series data often exhibits seasonality and trends that can make it challenging to model the underlying patterns. You can remove these components using techniques like differencing, where you compute the difference between consecutive observations, or by applying decomposition methods like Seasonal and Trend decomposition using Loess (STL) or seasonal differencing.

- **Generating Lag Features:** Lag features are created by shifting the original time series data by one or more time steps. These features help capture temporal dependencies and can improve the performance of your time series models. For example, if you are predicting stock prices, you might create lag features for the previous day's price, as well as prices from a week or a month ago.

- **Feature Scaling:** Feature scaling is important for time series data, especially when working with algorithms that are sensitive to the scale of input features, such as neural networks or support vector machines. Common feature scaling methods include normalization (scaling features to a range of [0, 1]) and standardization (scaling features to have zero mean and unit variance).

Practical Example

Imagine you have a dataset containing daily weather data, and you want to build a model to predict the temperature for the next day. To preprocess the time series data, you would:

1. Handle any missing values in the dataset using forward filling, linear interpolation, or a more advanced method that maintains the temporal structure.

2. Generate lag features to capture the previous day's temperature, as well as temperature values from a week or a month ago, to include relevant temporal information in your model.

3. Remove seasonality and trends, if present, using differencing or decomposition methods to ensure your model focuses on the underlying patterns in the data.

4. Resample the dataset to the desired frequency (e.g., daily, weekly, or monthly) based on your analysis requirements.

5. Scale the features using normalization or standardization to ensure the input features are on a consistent scale.

With the preprocessed time series data, you can now train and evaluate time series models, such as autoregressive integrated moving average (ARIMA), exponential smoothing state space models, or deep learning models like Long Short-Term Memory (LSTM) networks or transformers.

Performing these time series preprocessing steps will allow you to improve the performance and effectiveness of your time series models, ensuring they accurately capture the temporal patterns and relationships within the data.

Feature Scaling for Specific Algorithms

Feature scaling is an essential preprocessing step for many machine learning algorithms, as it ensures that input features are on a consistent scale, leading to better model performance and faster convergence. In this section, we will discuss the importance of feature scaling for specific algorithms and provide practical examples to demonstrate its impact.

Feature Scaling Methods

- **Normalization (Min-Max Scaling):** This method scales features to a specified range, typically [0, 1]. To normalize a feature, you subtract the minimum value and divide by the range (maximum value minus minimum value).

- **Standardization (Z-score Scaling):** This method scales features to have zero mean and unit variance. To standardize a feature, you subtract the mean and divide by the standard deviation.

Feature Scaling for Specific Algorithms

- **K-Nearest Neighbors (KNN):** KNN is a distance-based algorithm that relies on calculating the distance between data points to make predictions. If features are on different scales, larger-scaled features will dominate the distance calculation, leading to biased predictions. Feature scaling ensures that all features contribute equally to the distance calculation.

- **Support Vector Machines (SVM):** SVM aims to find the optimal hyperplane that separates classes by maximizing the margin. When features are on different scales, the algorithm may be sensitive to the scale of input features, leading to suboptimal hyperplanes. Feature scaling ensures that the algorithm can find the best hyperplane.

- **Neural Networks:** Neural networks use gradient-based optimization techniques, such as stochastic gradient descent (SGD), to minimize the loss function. When features are on different scales, the loss function's shape can be elongated, leading to slower convergence and a higher likelihood of getting stuck in local minima. Feature scaling ensures faster convergence and improved model performance.

- **Principal Component Analysis (PCA):** PCA is a dimensionality reduction technique that projects data onto the directions of maximum variance. When features are on different scales, the directions with higher variance will dominate the projection, leading to a biased representation. Feature scaling ensures that all features contribute equally to the PCA projection.

Practical Example

Imagine you have a dataset containing features related to customer transactions, such as transaction amount, transaction frequency, and customer age. Transaction amounts are in the thousands, while transaction frequency and customer age are in smaller units. You want to build a KNN model to predict customer churn based on these features.

To preprocess the data and apply feature scaling, you would:

1. Determine the appropriate feature scaling method for your data. In this case, you could use standardization, as it is less sensitive to outliers and works well for most machine learning algorithms.

2. Calculate the mean and standard deviation for each feature.

3. Subtract the mean and divide by the standard deviation for each feature to standardize the values.

4. Train your KNN model on the standardized data.

Applying feature scaling will allow you to ensure that all input features contribute equally to the distance calculation in the KNN algorithm, leading to more accurate and unbiased predictions. Regardless of the specific algorithm you are using, feature scaling is an essential preprocessing step that can significantly improve the performance of your machine learning models.

Advanced Feature Engineering Techniques

Feature engineering is the process of creating new features or transforming existing ones to improve the performance of your machine learning models. Advanced feature engineering techniques can help you capture complex patterns and relationships in your data that may not be apparent using basic transformations. In this section, we will discuss several advanced feature engineering techniques and provide practical examples to demonstrate their impact.

Advanced Feature Engineering Techniques

- **Interaction Features:** Interaction features are created by combining two or more existing features, often by applying arithmetic operations like multiplication, addition, or division. These features can help capture the combined effect of multiple features on the target variable.

- **Polynomial Features:** Polynomial features are created by raising existing features to a power or combining features using polynomial terms. These features can help capture non-linear relationships in your data and improve the performance of linear models.

- **Feature Selection:** Feature selection involves identifying the most important features for your model, allowing you to reduce dimensionality and improve model performance. Techniques like recursive feature elimination (RFE), LASSO regularization, or feature importance from tree-based models can help you identify the most relevant features.

- **Feature Encoding:** Feature encoding is the process of converting categorical variables into numerical representations suitable for machine learning algorithms. Advanced encoding techniques like target encoding or mean encoding can help capture the relationship between categorical variables and the target variable.

- **Domain-Specific Features:** Domain-specific features are created based on domain knowledge or expert insights about the problem at hand. These features can help capture unique relationships that may not be apparent using generic feature engineering techniques.

Practical Example

Imagine you have a dataset containing features related to real estate properties, such as the number of bedrooms, square footage, and neighborhood. You want to build a model to predict the property's sale price based on these features. To apply advanced feature engineering techniques, you would:

- Create interaction features, such as the product of the number of bedrooms and square footage, to capture the combined effect of these features on the sale price.

- Generate polynomial features, such as the square of the square footage or the interaction between square footage and the number of bedrooms, to capture non-linear relationships in the data.

- Perform feature selection using methods like RFE or LASSO regularization to identify the most important features for predicting the sale price.

- Encode categorical features, such as the neighborhood, using advanced encoding techniques like target encoding or mean encoding, to capture the relationship between the neighborhood and the sale price.

- Incorporate domain-specific features, such as the property's proximity to schools, parks, or public transportation, based on expert knowledge of the real estate market.

Applying these advanced feature engineering techniques will allow you to create a richer and more informative set of features for your machine learning models, leading to improved performance and more accurate predictions. Keep in mind that feature engineering is an iterative process, and it is essential to continually experiment with different techniques and evaluate their impact on your models.

Module 8 Exam Questions

1. What is the primary goal of data preprocessing in data science?

a) Building complex models

b) Visualizing data

c) Cleaning and transforming data to improve model performance

d) Increasing the amount of data

2. Which Python library is primarily used for data manipulation and analysis?

a) NumPy

b) Pandas

c) Matplotlib

d) Seaborn

3. Which of the following is a popular supervised machine learning task?

a) Clustering

b) Dimensionality reduction

c) Regression

d) Association rule learning

4. What is the primary purpose of feature scaling?

a) To create new features

b) To ensure input features are on a consistent scale

c) To reduce the number of features

d) To transform categorical features into numerical values

5. Which of the following is NOT a method of handling missing values in a dataset?

a) Forward filling

b) Backward filling

c) Linear interpolation

d) Maximum filling

6. In the context of time series data, what is resampling?

a) Changing the order of data points

b) Changing the frequency of the data points

c) Removing trends and seasonality

d) Generating lag features

7. Which feature scaling method scales features to have zero mean and unit variance?

a) Normalization

b) Min-Max Scaling

c) Standardization

d) None of the above

8. Which machine learning algorithm is sensitive to the scale of input features and often requires feature scaling?

a) Decision Trees

b) K-Nearest Neighbors

c) Naive Bayes

d) None of the above

9. In a time series dataset, which technique can be used to remove seasonality?

a) Differencing

b) Resampling

c) Forward filling

d) Backward filling

10. What is the primary purpose of exploratory data analysis with visualizations?

a) To create a final report

b) To understand the underlying structure and relationships in the data

c) To preprocess the data

d) To build machine learning models

11. What is the main purpose of feature engineering?

a) To reduce the amount of data

b) To visualize the data

c) To create new features or transform existing ones to improve model performance

d) To encode categorical variables

12. Which of the following is a technique for handling imbalanced data?

a) Standardization

b) Oversampling

c) Differencing

d) Mean encoding

13. In a machine learning context, what does PCA stand for?

a) Polynomial Component Analysis

b) Principal Component Analysis

c) Partial Correlation Analysis

d) Predictive Component Analysis

14. What is a typical use case for unsupervised machine learning algorithms?

a) Classification

b) Regression

c) Clustering

d) None of the above

15. Which of the following Python libraries is primarily used for data visualization?

a) NumPy

b) Pandas

c) Matplotlib

d) Scikit-learn

16. Which of the following is an advanced feature engineering technique?

a) Feature Scaling

b) Handling Missing Values

c) Interaction Features

d) Resampling

17. What is the primary goal of data cleaning?

a) To visualize the data

b) To build machine learning models

c) To identify and correct errors, inconsistencies, and inaccuracies in the data

d) To create new features

18. What is the primary purpose of creating lag features in time series data?

a) To remove trends and seasonality

b) To capture temporal dependencies

c) To handle missing values

d) To resample the data

19. Which of the following is a common method for encoding categorical variables?

a) Standardization

b) One-hot encoding

c) Differencing

d) Resampling

20. In the context of text data preprocessing, what does tokenization refer to?

a) Removing stop words

b) Converting text to lower case

c) Splitting text into individual words or phrases

d) Stemming or lemmatization

Answers:

1. c
2. b
3. c
4. b
5. d
6. b
7. c
8. b
9. a
10. b
11. c
12. b
13. b
14. c
15. c
16. c
17. c
18. b
19. b
20. c

Module 9. Data Visualization Techniques

Introduction to Data Visualization Techniques

Welcome to Module 9 on Data Visualization Techniques. In this module, you will learn the importance of effectively presenting data through visualizations and how to create impactful visual representations of your findings. Data visualization is a critical skill in data science, as it enables you to communicate complex information in a manner that is both engaging and easy to understand. By mastering these techniques, you will be able to make more informed decisions and better convey your insights to others.

We will begin this module by exploring the fundamental principles of data visualization and how they apply to various types of data. You will learn how to choose the right visualization methods depending on your data and the message you wish to convey. Throughout this module, we will discuss a range of practical examples that demonstrate the power of effective data visualization. These examples will showcase how visual representations can reveal patterns, trends, and relationships that may not be immediately apparent from the raw data alone.

Next, we will delve into the art of storytelling with data. You will discover how to create a compelling narrative around your data visualizations, making them even more engaging and meaningful. By combining powerful visual elements with a well-crafted story, you can captivate your audience and ensure that your insights are not only understood but also remembered.

Finally, we will explore the tools and techniques available for creating visually appealing and informative data visualizations. You will learn how to utilize various software packages and programming languages to develop interactive, dynamic, and aesthetically pleasing visualizations that effectively communicate your findings.

By the end of this module, you will have gained a solid understanding of the principles and techniques behind data visualization, allowing you to transform raw data into insightful and impactful visual stories. Armed with this knowledge, you will be well-equipped to present your data-driven insights to colleagues, stakeholders, and decision-makers in a clear and compelling manner.

Exploratory Data Analysis with Visualizations

In this section, we will delve into exploratory data analysis (EDA) using visualizations. EDA is a critical step in the data science process, as it allows you to gain a better understanding of your dataset and uncover hidden patterns, trends, and anomalies. By incorporating visualizations into your EDA, you can more effectively explore your data, identify relationships between variables, and generate hypotheses for further analysis.

To help you become proficient in performing EDA with visualizations, we will discuss several key techniques and provide practical examples to illustrate their use. By the end of this section, you will be able to apply these methods to your own data and gain valuable insights that can inform your decision-making.

- **Univariate analysis:** The first step in EDA is to examine each variable individually. Visualizations such as histograms, bar charts, and box plots can be used to study the distribution, central tendency, and dispersion of a single variable. For example, you may use a histogram to visualize the frequency distribution of a continuous variable like product sales, identifying trends or outliers that require further investigation.

- **Bivariate analysis:** After examining individual variables, the next step is to explore relationships between pairs of variables. Scatter plots, line charts, and heat maps can help you identify correlations, trends, or other interesting patterns between two variables. For example, you might create a scatter plot to visualize the relationship between advertising spend and sales revenue, enabling you to assess whether increased advertising is associated with higher sales.

- **Multivariate analysis:** In some cases, you may want to investigate relationships among multiple variables simultaneously. Techniques such as parallel coordinates plots, treemaps, and bubble charts can help you visualize complex, high-dimensional data. For instance, a bubble chart could be used to visualize the relationship between three variables like product category, sales revenue, and profit margin, revealing which categories contribute the most to your overall profitability.
- **Time series analysis:** If your data contains a temporal component, visualizations such as line charts, bar charts, and area charts can help you examine trends, seasonality, and cyclical patterns over time. For example, you might use a line chart to visualize monthly sales data over a two-year period, enabling you to identify seasonal fluctuations and longer-term trends that can inform your sales forecasts.
- **Geospatial analysis:** For datasets with geographical information, maps and spatial visualizations can provide valuable insights into spatial patterns, trends, and relationships. Choropleth maps, heat maps, and dot maps are just a few examples of geospatial visualizations that can help you explore your data. For instance, a choropleth map could be used to visualize the distribution of income levels across different regions, highlighting areas of high or low income concentration.

To get the most out of your EDA with visualizations, consider the following best practices:

- **Choose the appropriate visualization type:** Select a visualization method that is well-suited to the data you are working with and the insights you are trying to uncover. Experiment with different visualizations to find the most effective way to communicate your findings.

- **Use clear and consistent formatting:** Ensure that your visualizations are easy to interpret by using clear labels, legends, and scales. Consistent formatting across multiple visualizations can also make it easier for your audience to understand and compare different aspects of your data.

- **Highlight key insights:** Draw attention to important findings or patterns in your visualizations by using colors, annotations, or other visual cues. This can help guide your audience's focus to the most relevant aspects of your data and support the narrative you are trying to convey.

- **Interactivity:** Incorporate interactive elements in your visualizations when possible, allowing users to engage with your data and explore it more deeply. Interactive visualizations can enable your audience to filter, zoom, or drill down into specific areas of interest, leading to a more personalized and immersive experience.

- **Iterate and refine:** EDA is an iterative process, and so is creating effective visualizations. As you explore your data, you may find that certain visualizations reveal new questions or highlight unexpected patterns. Be prepared to revise your visualizations as you uncover new insights and refine your understanding of the dataset.

To perform EDA with visualizations, you can leverage various tools and libraries, such as Matplotlib, Seaborn, and Plotly in Python or ggplot2 and Shiny in R. These tools offer a wide range of visualization options and customization features, empowering you to create compelling visual representations of your data that can enhance your EDA and inform your decision-making.

In summary, incorporating visualizations into your exploratory data analysis can significantly improve your ability to understand and communicate the patterns, trends, and relationships within your dataset. By mastering the techniques and best practices discussed in this section, you will be well-equipped to perform insightful EDA with visualizations and uncover valuable insights that can guide your data-driven decision-making.

Creating Effective Visualizations

In this section, we will focus on the principles and techniques for creating effective visualizations that communicate your data-driven insights clearly and concisely. By adhering to these guidelines, you can ensure that your visualizations not only look visually appealing but also successfully convey your message and engage your audience. We will provide practical examples and tips to help you develop impactful visualizations that resonate with your viewers.

- **Choose the right visualization type:** Selecting the appropriate visualization method for your data is critical to conveying your message effectively. Consider the nature of your data, the relationships you want to emphasize, and the insights you want to communicate. For example, use bar charts for categorical data, line charts for time series data, and scatter plots for examining relationships between two continuous variables.

- **Keep it simple and clean:** Cluttered and overly complex visualizations can be difficult to understand and may detract from your message. Simplify your visualizations by removing unnecessary elements, such as excessive gridlines, borders, or background colors. Use whitespace effectively to create a clean, uncluttered design that allows your data to stand out.

- **Use color strategically:** Color can be a powerful tool for highlighting key insights, differentiating data categories, and improving the overall aesthetics of your visualization. However, it's essential to use color judiciously. Choose a color palette that is visually appealing, easy to interpret, and accessible to viewers with color vision deficiencies. Be consistent in your use of color across multiple visualizations to maintain coherence and improve comprehension.

- **Optimize text elements:** Clear and concise labels, titles, and annotations are crucial for helping your audience interpret your visualizations. Use descriptive titles that summarize the main insights, label axes and data points accurately, and add annotations to highlight significant findings or patterns. Choose a legible font and ensure that text elements are sized appropriately for readability.

- **Maintain aspect ratio:** An appropriate aspect ratio is essential for maintaining the integrity of your data representation. A distorted aspect ratio can lead to misinterpretations and inaccurate conclusions. Adjust the aspect ratio of your visualization to ensure that your data is displayed accurately and consistently across various devices and screen sizes.

- **Use interactivity wisely:** Interactive visualizations can enhance user engagement and provide a more immersive experience. However, it's important not to overuse interactivity or create visualizations that are overly complex. Implement interactivity when it adds value, such as allowing users to filter data, zoom in on specific data points, or explore relationships between variables in more detail.

- **Tell a story with your data:** Effective visualizations are not just about presenting data but also about crafting a narrative that resonates with your audience. Organize your visualizations in a logical sequence that builds on the insights and relationships you have uncovered during your analysis. Use clear, compelling language and visual elements to guide your audience through your data story, helping them understand the context and relevance of your findings.

- **Test and refine:** Creating effective visualizations is an iterative process. Be prepared to test your visualizations with a diverse group of users, gather feedback, and make refinements based on their input. Continuously refining your visualizations will ensure that they are as clear, engaging, and informative as possible.

Following these principles and techniques will allow you to create effective visualizations that not only look visually appealing but also successfully communicate your insights and engage your audience. Remember to be mindful of your audience's needs and preferences, and tailor your visualizations to suit their level of expertise and familiarity with the subject matter.

To further illustrate these principles, let's consider a practical example. Suppose you are analyzing the impact of a marketing campaign on monthly sales for a retail store. You want to create a visualization that clearly communicates the relationship between marketing spend and sales revenue.

- **Choose the right visualization type:** In this case, a scatter plot would be an appropriate choice, as it allows you to visualize the relationship between two continuous variables—marketing spend and sales revenue.

- **Keep it simple and clean:** Design a scatter plot with a clean background and minimal gridlines, so the focus remains on the data points representing the relationship between marketing spend and sales.

- **Use color strategically:** Apply a color scale that represents the density of data points or use different colors for pre- and post-campaign periods to highlight the campaign's impact on sales.

- **Optimize text elements:** Add a clear title, such as "Impact of Marketing Campaign on Monthly Sales," label the x and y-axes as "Marketing Spend" and "Sales Revenue," respectively, and include any necessary annotations to emphasize key findings.

- **Maintain aspect ratio:** Ensure that the aspect ratio of your scatter plot accurately represents the data and avoids distortion.
- **Use interactivity wisely:** If you decide to make your scatter plot interactive, you could allow users to filter the data by product category, view details of specific data points, or toggle between pre- and post-campaign data.
- **Tell a story with your data:** Present your scatter plot as part of a broader data story that includes additional visualizations and narrative elements to provide context, highlight key insights, and guide your audience through your analysis.
- **Test and refine: Share** your visualization with colleagues or stakeholders, gather feedback, and iterate on your design to ensure it effectively communicates your message and resonates with your audience.

Applying these principles and techniques to your own visualizations will allow you to be well-equipped to create compelling and effective visual representations of your data that help you communicate your insights and engage your audience.

Storytelling with Data

In this section, we will explore the art of storytelling with data, a powerful technique that enables you to communicate your data-driven insights more effectively and engage your audience. By combining compelling narratives with impactful visualizations, you can create a memorable experience that helps your audience understand and retain the information you present. We will provide detailed information and practical examples to help you become proficient in storytelling with data.

- **Understand your audience:** Before you begin crafting your data story, it's crucial to understand who your audience is, their background, and their needs. This will help you tailor your narrative and visualizations to suit their level of expertise, interests, and expectations. Consider the questions your audience may have and address them in your story.

- **Define your key message:** Identify the main insights or conclusions you want your audience to take away from your data story. This will help you focus your narrative and visualizations on the most relevant and impactful information. Be concise and clear in conveying your key message, making it easy for your audience to understand and remember.

- **Create a logical structure:** Organize your data story in a logical sequence that guides your audience through your analysis and builds towards your key message. Start with an introduction that sets the context and presents your objectives. Then, lead your audience through your data exploration and findings, using visualizations and narrative elements to highlight important insights and relationships. Finally, conclude with a summary that reinforces your key message and provides actionable recommendations or next steps.

- **Use visualizations effectively:** Select the appropriate visualization types to support your narrative and emphasize the insights you want to convey. Ensure that your visualizations are clear, easy to interpret, and visually appealing. Align your visualizations with the narrative elements of your data story, using annotations, color, and other visual cues to draw attention to key insights.

- **Employ narrative techniques:** Engage your audience with storytelling techniques, such as setting the context, creating tension or suspense, and using relatable examples or analogies. This will make your data story more engaging and memorable. Use clear, concise language that is accessible to your audience, avoiding jargon and technical terms that may be unfamiliar or confusing.

- **Be transparent and credible:** Ensure that your data story is based on accurate, reliable data and that your conclusions are supported by evidence. Be transparent about your data sources, methodology, and assumptions, as this will build trust with your audience and enhance the credibility of your insights.

- **Iterate and refine:** Creating a compelling data story is an iterative process. Gather feedback from your audience and refine your narrative and visualizations based on their input. Continuously improving your data story will ensure that it is as clear, engaging, and informative as possible.

To illustrate these principles in practice, let's consider an example. Suppose you are a data analyst at a healthcare organization, and you have been tasked with presenting an analysis of patient satisfaction survey results.

- **Understand your audience:** Your audience may include hospital administrators, medical professionals, and other stakeholders interested in improving patient satisfaction.

- **Define your key message:** Your key message could be identifying areas of improvement to enhance patient satisfaction and the overall patient experience.

- **Create a logical structure:** Start by introducing the objective of your analysis, then present the survey results, highlighting trends and patterns in patient satisfaction across various dimensions, such as department, age group, or type of service. Discuss the areas where patient satisfaction is particularly high or low, using visualizations to emphasize these findings. Conclude with recommendations for improvement and a call to action for your stakeholders.

- **Use visualizations effectively:** Create a series of visualizations that showcase your findings, such as bar charts to compare satisfaction scores across departments, heatmaps to visualize patterns in survey responses, or line charts to track satisfaction trends over time. Ensure that these visualizations are clear, visually appealing, and aligned with your narrative.

- **Employ narrative techniques:** Set the context by explaining the importance of patient satisfaction for the healthcare organization and its impact on patient outcomes. Create tension by highlighting areas where satisfaction scores are low and the need for improvement is urgent. Use relatable examples or anecdotes to illustrate the real-life experiences of patients and make your data story more engaging.

- Be transparent and credible: Clearly state your data sources, methodology, and any limitations of your analysis. Ensure that your conclusions are well-supported by the data, and be open about any uncertainties or assumptions.

- Iterate and refine: Share your data story with colleagues or stakeholders, gather feedback, and make refinements based on their input. Continuously improving your data story will ensure that it is as clear, engaging, and informative as possible.

Applying these principles and techniques to your own data storytelling efforts will allow you to create compelling narratives that not only inform but also engage and inspire your audience. Through effective storytelling with data, you can elevate your data-driven insights, making them more memorable and impactful, ultimately leading to better decision-making and improved outcomes.

Accessibility in Data Visualization

In this section, we will discuss the importance of accessibility in data visualization and provide practical guidelines for creating visualizations that can be understood and appreciated by a diverse audience, including people with disabilities such as color vision deficiencies or visual impairments. By making your visualizations more accessible, you ensure that your insights can reach a wider audience and promote a more inclusive data-driven culture.

- **Use color palettes that are colorblind-friendly:** Color vision deficiencies, such as red-green colorblindness, affect a significant portion of the population. To make your visualizations more accessible, use color palettes that are distinguishable by people with various types of color vision deficiencies. Tools like ColorBrewer or the viridis package in Python can help you choose appropriate color palettes.

- **Provide sufficient contrast:** Ensure that there is enough contrast between text, background, and other visual elements in your visualizations. High contrast helps those with visual impairments or low vision to better distinguish between different elements. Use online tools like the WebAIM Contrast Checker to evaluate and improve the contrast in your visualizations.

- **Use clear and legible text:** Choose fonts that are easy to read, and avoid using overly stylized or decorative typefaces. Ensure that your text size is large enough to be easily readable, and consider using bold or high-contrast text to emphasize important information. Additionally, use a consistent and clear hierarchy for text elements such as titles, labels, and annotations to guide your audience through the visualization.

- **Offer alternative text descriptions:** For users who rely on screen readers or other assistive technologies, provide alternative text descriptions for your visualizations. These descriptions should clearly and concisely convey the main insights and information presented in the visualization, allowing users to understand the content even if they cannot directly perceive the visual elements.

- **Ensure keyboard navigation:** Make sure your interactive visualizations are accessible through keyboard navigation, as some users may not be able to use a mouse or touch screen. Ensure that all interactive elements can be accessed, operated, and navigated using only a keyboard, and provide clear focus indicators to show which element is currently selected.
- **Avoid relying solely on color to convey information:** In addition to using color, use other visual cues such as shapes, patterns, or labels to distinguish between different data series or categories. This will make your visualizations more accessible to individuals with color vision deficiencies and also improve the overall clarity of your visualizations for all users.
- **Test your visualizations for accessibility:** Use accessibility testing tools, such as the Web Accessibility Evaluation Tool (WAVE) or the Chrome Accessibility Developer Tools, to identify and address any accessibility issues in your visualizations. Additionally, seek feedback from users with diverse abilities to ensure your visualizations are truly accessible and inclusive.

To illustrate these principles with a practical example, imagine you are creating a bar chart that compares the revenue of different product categories for an online store. To make this visualization more accessible, you could:

- Choose a colorblind-friendly color palette for the bars representing different product categories.

- Ensure there is sufficient contrast between the bar colors, text, and background.
- Use a clear, legible font for axis labels, titles, and annotations.
- Provide an alternative text description of the bar chart, summarizing the key findings and trends.
- If the bar chart is interactive, ensure that users can navigate and interact with the chart using only a keyboard.
- Use patterns or labels in addition to color to distinguish between product categories, if necessary.
- Test the final visualization for accessibility and gather feedback from users with diverse abilities.

Following these guidelines and incorporating accessibility best practices into your data visualization process will allow you to create more inclusive visualizations that effectively communicate your insights to a

broader audience. Not only does this promote a more inclusive data-driven culture, but it also enhances the overall clarity and effectiveness of your visualizations, making them more impactful for all users.

In summary, by considering accessibility throughout the design and creation of your data visualizations, you can ensure that your insights are more widely understood and appreciated. By implementing the practical guidelines and examples provided in this section, you can create visualizations that are more inclusive and engaging, ultimately making your data-driven insights more valuable and impactful for a diverse audience.

Advanced Visualization Techniques

In this section, we will explore advanced visualization techniques that can help you create more insightful and engaging visualizations for complex datasets. These techniques go beyond the basic charts and plots covered in previous sections, allowing you to better represent multidimensional, hierarchical, or interconnected data. We will provide detailed information and practical examples for each technique, helping you expand your visualization toolkit and create more impactful data-driven insights.

Treemaps

Treemaps are a space-filling visualization technique used to represent hierarchical data using nested rectangles. Each rectangle represents a data item, and its size corresponds to a quantitative variable, such as revenue or population. Treemaps are useful for visualizing large datasets with multiple levels of hierarchy and can help you identify patterns or outliers within the data.

Practical example: If you want to visualize the market share of different smartphone manufacturers and their individual models, you can create a treemap where each manufacturer is represented by a large rectangle, and each model is a smaller rectangle nested within the manufacturer's rectangle. The size of each rectangle can represent the number of units sold, allowing you to easily compare market shares across manufacturers and models.

Parallel Coordinates

Parallel coordinates are a technique for visualizing high-dimensional data using parallel, equally spaced vertical lines (axes) that represent different variables. Data points are plotted as connected line segments across these axes, allowing you to observe relationships between multiple variables simultaneously. Parallel coordinates are useful for identifying patterns, clusters, or outliers within high-dimensional datasets.

Practical example: If you are analyzing a dataset of cars with multiple attributes, such as price, horsepower, fuel efficiency, and weight, you can use parallel coordinates to visualize the relationships between these attributes. By connecting the data points across the axes, you can observe how different combinations of attributes relate to one another and identify potential clusters or outliers in the data.

Sankey Diagrams

Sankey diagrams are used to visualize the flow of data between different nodes or categories. They consist of connected flow lines, where the width of each line represents the quantity of the flow, such as energy, money, or resources. Sankey diagrams are useful for visualizing the distribution of data across multiple stages or categories, and for illustrating the flow of data in complex systems.

Practical example: If you want to visualize the flow of website traffic from different traffic sources (such as organic search, social media, or referrals) to various pages on your website and ultimately to conversion, you can use a Sankey diagram. This will help you identify the most effective traffic sources and better understand user navigation patterns, allowing you to optimize your website for improved conversion rates.

Network Graphs

Network graphs are used to visualize relationships between entities in a dataset, such as people, organizations, or concepts. Nodes represent the entities, and edges represent the connections or relationships between them. Network graphs can help you identify clusters, central nodes, or influential entities within a network, and can be used to analyze social networks, communication networks, or other interconnected systems.

Practical example: If you want to analyze the relationships between characters in a novel or a movie, you can create a network graph where each character is a node and each interaction between characters is an edge. This visualization can help you identify central characters, reveal the structure of relationships within the story, and provide insights into the narrative's complexity.

Choropleth Maps

Choropleth maps are used to visualize geographic data by shading regions or areas based on a quantitative variable, such as population density, income, or unemployment rate. Choropleth maps can help you identify spatial patterns and trends in your data and are commonly used in fields like public health, economics, or environmental science.

Practical example: If you want to visualize the distribution of median household income across different counties in a state, you can create a choropleth map where each county is shaded according to its median income level. This visualization can help you identify areas with high or low income levels and reveal potential correlations with other geographic or demographic factors.

Heatmaps

Heatmaps are used to visualize the density or intensity of data points in a two-dimensional space, such as geographic locations or a matrix of values. Data points are represented by color gradients, with different colors indicating different levels of density or intensity. Heatmaps can help you identify clusters or hotspots in your data and are commonly used in fields like biology, finance, or sports analytics.

Practical example: If you want to analyze the distribution of crime incidents in a city, you can create a heatmap where each incident is represented by a colored point on a map. The intensity of the color indicates the density of crime incidents in that area, allowing you to easily identify high-crime areas or potential patterns in the distribution of crime incidents.

Scatterplot Matrix

A scatterplot matrix is a grid of scatterplots, where each pair of variables in a dataset is plotted against each other. This visualization technique allows you to observe the relationships between multiple variables simultaneously and can help you identify patterns, correlations, or outliers in your data.

Practical example: If you are analyzing a dataset of housing prices with multiple attributes, such as square footage, number of bedrooms, age, and location, you can use a scatterplot matrix to visualize the relationships between these attributes. By examining the scatterplots, you can identify potential correlations between variables, such as a positive relationship between square footage and price, or an inverse relationship between age and price.

Incorporating these advanced visualization techniques into your data analysis will allow you to create more insightful and engaging visualizations that effectively communicate complex data-driven insights. These techniques allow you to represent multidimensional, hierarchical, or interconnected data in a visually appealing and informative way, ultimately enhancing the impact and value of your data visualizations.

Geospatial Data Visualization

In this section, we will explore geospatial data visualization techniques that allow you to create insightful and engaging visualizations using geographic or location-based data. Geospatial data visualization can help you reveal patterns, trends, and relationships in your data that are tied to geographic locations or regions. We will provide detailed information and practical examples for each technique, helping you better understand and effectively communicate spatial information in your data.

Point Maps

Point maps are used to visualize individual data points on a map, with each point representing a specific location, such as an address, a city, or a landmark. Point maps are useful for visualizing the distribution of events, locations, or other geographically anchored data points.

Practical example: If you want to visualize the locations of all the restaurants in a city, you can create a point map where each point represents a restaurant. This visualization can help you identify clusters of restaurants in particular neighborhoods or highlight areas with fewer dining options.

Heatmaps

As mentioned earlier, heatmaps can also be used for geospatial data visualization. In the context of geospatial data, heatmaps represent the density or intensity of data points in a geographic space, with different colors indicating different levels of density.

Practical example: If you want to visualize the distribution of crime incidents in a city, you can create a heatmap where each incident is represented by a colored point on a map. The intensity of the color indicates the density of crime incidents in that area, allowing you to easily identify high-crime areas or potential patterns in the distribution of crime incidents.

Choropleth Maps

As previously mentioned, choropleth maps are used to visualize geographic data by shading regions or areas based on a quantitative variable. Choropleth maps can help you identify spatial patterns and trends in your data.

Practical example: If you want to visualize the distribution of median household income across different counties in a state, you can create a choropleth map where each county is shaded according to its median income level. This visualization can help you identify areas with high or low income levels and reveal potential correlations with other geographic or demographic factors.

Symbol Maps

Symbol maps use symbols, such as circles, squares, or other shapes, to represent data points on a map. The size or color of the symbols can be used to encode additional quantitative variables, such as population or revenue.

Practical example: If you want to visualize the population of cities in a country, you can create a symbol map where each city is represented by a circle. The size of each circle can be proportional to the city's population, allowing you to easily compare populations across cities and identify major urban centers.

Flow Maps

Flow maps are used to visualize the movement of data between geographic locations, such as migration patterns, trade flows, or transportation networks. Flow maps use lines or arrows to represent the direction and magnitude of the flow, with the thickness or color of the lines encoding the quantity or intensity of the movement.

Practical example: If you want to visualize the flow of commuters between different neighborhoods in a city, you can create a flow map where each neighborhood is represented by a node, and the flow of commuters is represented by lines connecting the nodes. The thickness of the lines can represent the number of commuters, allowing you to easily identify major commuting routes and patterns in the city.

Cartograms

Cartograms are a type of map that distorts the size or shape of geographic regions to represent a quantitative variable, such as population or GDP. Cartograms can help you emphasize the importance of certain regions or reveal patterns and relationships in your data that are not apparent in traditional maps.

Practical example: If you want to visualize the distribution of global GDP by country, you can create a cartogram where each country is resized according to its GDP. This visualization can help you identify countries with high or low GDP and reveal potential correlations with other geographic or demographic factors.

3D Maps

3D maps are used to visualize geographic data in three dimensions, adding an additional layer of depth to your visualizations. 3D maps can be used to represent elevation, building heights, or other z-axis information, and can provide a more immersive and engaging experience for your audience.

Practical example: If you want to visualize the elevation of a mountain range, you can create a 3D map that shows the terrain in three dimensions. This visualization can help you better understand the topography of the area and identify features such as peaks, valleys, or plateaus.

Geospatial Data Animation

Geospatial data animation is the process of visualizing geographic data over time, allowing you to observe patterns, trends, or changes in your data as they occur. Animated maps can be particularly effective for visualizing time-series data or data that changes over time, such as population growth, climate change, or the spread of a disease.

Practical example: If you want to visualize the spread of a disease over time, you can create an animated map that shows the number of cases in different regions as the disease progresses. This visualization can help you identify patterns in the spread of the disease and inform strategies for containment or mitigation.

Incorporating these geospatial data visualization techniques into your data analysis will allow you to create more insightful and engaging visualizations that effectively communicate location-based information in your data. These techniques allow you to represent geographic data in a visually appealing and informative way, ultimately enhancing the impact and value of your data visualizations. Whether you are working with location-based data for the first time or looking to expand your geospatial visualization skills, these techniques will provide you with a solid foundation for creating effective and impactful geospatial data visualizations.

Dashboard Design and Best Practices

In this section, we will discuss dashboard design and best practices, focusing on how to create effective, user-friendly, and impactful dashboards for your data visualizations. Dashboards are powerful tools for consolidating and presenting complex data in a visually appealing and easily digestible format. By following best practices in dashboard design, you can ensure that your audience can quickly understand the insights you want to communicate, leading to better decision-making and data-driven actions. We will provide detailed information and practical examples to help you create effective dashboards.

Identify your audience and their needs

Before you start designing your dashboard, it's essential to understand who your audience is and what information they need. Consider their level of expertise, their goals, and the key questions they need to answer using your dashboard. This will help you determine the most relevant data and visualizations to include.

Practical example: If you are creating a dashboard for sales managers, you might want to focus on visualizing sales performance metrics, such as revenue, profit margins, and sales growth, as well as comparing performance across different regions or product lines.

Choose the right visualizations

Select the most appropriate visualization types for your data and audience, considering factors such as data complexity, relationships between variables, and the level of detail your audience needs. Make sure to use clear and consistent visuals that are easy to understand and interpret.

Practical example: If you want to show the relationship between two variables, such as sales and marketing spend, a scatter plot might be the most effective visualization. For displaying the distribution of a single variable, such as customer satisfaction scores, a histogram or box plot may be more appropriate.

Organize your dashboard logically

Arrange the visualizations and components on your dashboard in a way that is easy to navigate and understand. Group related visuals together, use a clear hierarchy to guide the user's eye, and provide ample white space to avoid clutter.

Practical example: You can organize your dashboard into sections or panels, with each section dedicated to a specific aspect of your data, such as sales performance, customer demographics, or product performance. This will help your audience easily locate the information they need and understand the relationships between different visuals.

Use consistent design elements

Ensure that your dashboard design is visually cohesive by using consistent colors, fonts, and formatting throughout. This will not only make your dashboard more aesthetically pleasing but also improve its readability and usability.

Practical example: Choose a color palette that complements your brand or theme and use it consistently across all visualizations. Similarly, select a font and font size that is easy to read and use it for all text elements on your dashboard.

Make your dashboard interactive

Including interactive elements in your dashboard, such as filters, drop-down menus, and sliders, can greatly enhance its usability and provide your audience with more control over the data they view.

Practical example: Add filters to your dashboard that allow users to view data for specific time periods, regions, or product categories.

This will enable them to focus on the data that is most relevant to their needs and explore different aspects of the data more easily.

Optimize for different devices and screen sizes

Ensure that your dashboard is accessible and usable on various devices, such as desktops, laptops, tablets, and smartphones. This will improve the user experience and make your dashboard more versatile.

Practical example: Use responsive design techniques to ensure that your dashboard automatically adjusts to fit different screen sizes and orientations. Test your dashboard on multiple devices to ensure that it remains functional and visually appealing across all platforms.

Test and iterate

Continuously refine and improve your dashboard based on user feedback and analytics. This will ensure that your dashboard remains relevant, useful, and effective over time.

Practical example: After launching your dashboard, gather feedback from your audience regarding its usability, design, and content. Monitor usage analytics to identify which visualizations are most frequently viewed or interacted with, and use this information to guide future improvements. Iterate on your design by making adjustments and enhancements based on this feedback and data.

Provide clear instructions and documentation

To help your audience get the most value from your dashboard, provide clear instructions on how to use it, along with any necessary documentation or resources. This will ensure that users can easily navigate and understand the dashboard, even if they are not familiar with the data or visualizations.

Practical example: Include a brief tutorial or guide within your dashboard that explains how to use the interactive elements, such as filters or dropdown menus. Provide additional resources or documentation, such as a glossary of terms or a detailed explanation of the data sources, to help users better understand the context and meaning of the visualizations.

Focus on performance

Optimize your dashboard's performance to ensure that it loads quickly and runs smoothly. Slow or unresponsive dashboards can frustrate users and detract from the overall user experience.

Practical example: Optimize your data queries to reduce the time it takes to load your dashboard. Consider using techniques such as data aggregation or pre-calculation to minimize the processing required when users interact with the dashboard. Additionally, avoid using too many high-resolution images or other resource-intensive elements that could slow down your dashboard's performance.

Tell a story with your data

Finally, use your dashboard to tell a compelling story with your data. This can help your audience better understand the context, meaning, and relevance of your visualizations, ultimately leading to more informed decision-making and action.

Practical example: Arrange your dashboard in a way that guides users through the data, highlighting key insights, trends, and patterns along the way. Use annotations, tooltips, or other explanatory elements to provide context and explanation for your visualizations, helping users understand the significance of the data and how it relates to their goals or objectives.

Following these dashboard design best practices will allow you to create visually appealing, user-friendly, and impactful dashboards that effectively communicate your data insights to your audience. As you continue to develop your skills in data visualization and dashboard design, remember that the ultimate goal is to create clear, engaging, and informative visualizations that empower your audience to make data-driven decisions and take meaningful action.

Data Visualization Tools and Libraries

In this section, we will explore various data visualization tools and libraries that can help you create impactful visualizations and dashboards. The choice of tool or library depends on your specific needs, technical skills, and the type of data you are working with. We will provide detailed information on each tool or library and practical examples to help you choose the right one for your projects.

Microsoft Excel

Excel is a widely-used spreadsheet application that offers various data visualization capabilities, including charts, graphs, and pivot tables. Excel is user-friendly and suitable for beginners, as well as those with limited programming skills.

Practical example: You can use Excel to create a simple bar chart comparing the sales performance of different products or a line chart to visualize the trend of monthly revenue.

Tableau

Tableau is a powerful data visualization tool that allows you to create interactive and shareable dashboards. With its drag-and-drop interface, Tableau is user-friendly and suitable for both beginners and advanced users. It can connect to various data sources, including spreadsheets, databases, and cloud-based storage.

Practical example: You can use Tableau to create an interactive dashboard that visualizes customer demographics, sales performance, and product trends, allowing your team to explore the data and uncover insights.

Power BI

Power BI is a business analytics service from Microsoft that enables you to create interactive visualizations and dashboards. It offers seamless integration with other Microsoft products, such as Excel and Azure, and provides a wide range of data connectors.

Practical example: You can use Power BI to create a dashboard that tracks key performance indicators (KPIs) for your business, such as revenue, customer acquisition cost, and employee productivity.

D3.js

D3.js is a JavaScript library for creating dynamic, interactive data visualizations in web browsers. It allows you to build custom visualizations using web standards such as HTML, SVG, and CSS. D3.js requires programming skills and is suitable for those with a background in web development.

Practical example: You can use D3.js to create a custom interactive map that visualizes the distribution of your customers or a dynamic scatter plot that updates in real-time as new data is added.

Python Libraries

Python offers several libraries for data visualization, including Matplotlib, Seaborn, Plotly, and Bokeh. These libraries enable you to create a wide range of static, interactive, and web-based visualizations using the Python programming language.

Practical example: You can use Matplotlib and Seaborn to create a histogram that shows the distribution of customer ages or use Plotly to create an interactive line chart that displays the growth of your user base over time.

R Libraries

R is a popular programming language for data analysis and visualization. It provides a variety of libraries for creating visualizations, such as ggplot2, Shiny, and Plotly for R.

Practical example: You can use ggplot2 to create a box plot that visualizes the spread of customer satisfaction scores or use Shiny to create a web-based dashboard that allows users to interact with your data and explore different visualizations.

QlikView/Qlik Sense

Qlik offers two data visualization and analytics tools: QlikView and Qlik Sense. Both provide powerful visualization capabilities and enable users to create interactive dashboards. QlikView is more focused on guided analytics, while Qlik Sense emphasizes self-service and ease of use with a drag-and-drop interface.

Practical example: You can use Qlik Sense to create a dashboard that visualizes the performance of different marketing channels, allowing users to interact with the data and analyze the return on investment (ROI) for each channel.

Google Data Studio

Google Data Studio is a free, web-based data visualization and reporting tool that allows you to create interactive, shareable dashboards. It offers seamless integration with other Google products, such as Google Analytics, Google Sheets, and Google BigQuery.

Practical example: You can use Google Data Studio to create a dashboard that visualizes your website's traffic, user engagement, and conversion metrics, making it easy to monitor your online marketing efforts and make data-driven decisions.

Looker

Looker is a data platform that offers powerful data visualization and exploration capabilities. It enables users to create custom visualizations, interactive dashboards, and reports using a web-based interface. Looker is particularly suited for organizations that require advanced data modeling and integration with their existing data infrastructure.

Practical example: You can use Looker to create a dashboard that visualizes key financial metrics, such as revenue, expenses, and net income, allowing your finance team to monitor the company's financial performance and identify trends.

Highcharts

Highcharts is a JavaScript charting library that allows you to create interactive, responsive, and highly customizable charts for web and mobile applications. It supports a wide range of chart types, including line, area, bar, pie, and scatter charts, as well as more advanced visualizations like heatmaps and treemaps.

Practical example: You can use Highcharts to create a responsive, interactive bar chart that compares the revenue of different products across various regions, enabling users to explore the data and identify patterns on different devices.

These data visualization tools and libraries offer a range of capabilities and features, catering to different skill levels, data types, and visualization requirements. As you explore these options, consider your specific needs, technical expertise, and the kind of visualizations you want to create. Choosing the right tool or library will allow you to enhance your data visualization skills and create impactful, engaging visuals that effectively communicate your insights and tell compelling stories with your data.

Module 9 Exam Questions

1. What does cybersecurity refer to?

a) The practice of protecting computers, networks, applications, and data from unauthorized access, theft, or damage

b) The process of creating secure passwords

c) The use of firewalls and antivirus software

d) The study of cybercriminals and their motives

1. What is the primary goal of data visualization?

a) To create visually appealing graphics

b) To effectively communicate data insights

c) To manipulate data

d) To store data

Which of the following is NOT a basic chart type?

a) Bar chart

b) Line chart

c) Pie chart

d) Word cloud

In which Python library can you create a histogram to visualize the distribution of a dataset?

a) Matplotlib

b) Seaborn

c) D3.js

d) QlikView

Which of the following tools is specifically designed for creating interactive dashboards?

a) Microsoft Excel

b) Tableau

c) D3.js

d) Matplotlib

What is the purpose of exploratory data analysis (EDA)?

a) To clean data

b) To perform feature engineering

c) To generate initial insights and patterns in the data

d) To develop machine learning models

Which data visualization tool is web-based and offers seamless integration with other Google products?

a) Tableau

b) Power BI

c) Google Data Studio

d) Qlik Sense

What is the main advantage of using geospatial data visualization?

a) To analyze data over time

b) To visualize data in 3D

c) To display data on a map and analyze spatial relationships

d) To create interactive charts

Which of the following techniques can be used to make your data visualizations more accessible?

a) Using high contrast colors

b) Removing all text from the visualization

c) Using only one color for all elements

d) Creating complex visualizations

What type of machine learning algorithm is best suited for visualizing clusters within a dataset?

a) Regression

b) Classification

c) Clustering

d) Dimensionality reduction

In the context of data visualization, what is the purpose of using tooltips?

a) To add interactivity to a chart

b) To provide additional information about data points

c) To change the chart type

d) To create animations

Which of the following libraries is used for creating interactive data visualizations in web browsers using JavaScript?

a) Matplotlib

b) Seaborn

c) D3.js

d) ggplot2

Which R library is commonly used to create high-quality static graphics?

a) Shiny

b) Plotly

c) D3.js

d) ggplot2

What is the primary purpose of dashboard design best practices?

a) To create visually appealing dashboards

b) To improve dashboard performance

c) To make dashboards more user-friendly

d) All of the above

What is the main benefit of using a line chart over a bar chart?

a) To show part-to-whole relationships

b) To visualize trends over time

c) To compare categories

d) To display hierarchical data

In which of the following scenarios would you most likely use a pie chart?

a) To display the distribution of ages in a population

b) To show the relationship between two continuous variables

c) To compare the market share of different companies

d) To visualize the number of sales per month

What is an important consideration when choosing colors for your data visualization?

a) Color contrast

b) The number of colors

c) Personal preferences

d) The shape of the data

Which of the following is an important aspect of storytelling with data?

a) Ensuring that data is accurate

b) Establishing a clear narrative

c) Using interactive visualizations

d) Incorporating animation

What is a key consideration when selecting a data visualization tool or library?

a) The size of the tool or library

b) The programming language

c) The compatibility with your data source

d) The level of interactivity it provides

When creating a bar chart, which of the following is true about the axis labels?

a) The x-axis should display categorical data

b) The y-axis should display categorical data

c) The x-axis should display continuous data

d) The y-axis should display hierarchical data

Which of the following is a common use case for a heat map?

a) To visualize the correlation between multiple variables

b) To display the distribution of data points on a map

c) To compare the performance of different categories

d) To display the relationship between two continuous variables

Answers:

1. b
2. d
3. a
4. b
5. c
6. c
7. c
8. a
9. c
10. b
11. c
12. d
13. d
14. b
15. c
16. a
17. b
18. c
19. a
20. a

Module 10. Building a Career in Data Science

Introduction to Building a Career in Data Science

Welcome to Module 10: Final Thoughts and Next Steps in Data Science. In this module, we will focus on the essential aspects of growing and nurturing your career in the ever-evolving world of data science. We will delve into practical strategies for showcasing your skills, staying current with industry trends, and fostering a sense of community with fellow professionals.

As you progress in your data science journey, it is crucial to build a robust portfolio that demonstrates your expertise and accomplishments. We will discuss the importance of curating projects that highlight your strengths, as well as how to present them effectively to potential employers and collaborators.

In addition to showcasing your work, it is vital to stay informed about the latest advancements and trends in the field of data science. We will explore various resources and strategies for staying up-to-date with cutting-edge techniques and tools, ensuring that you remain competitive in the job market and can adapt to new challenges.

Moreover, we will discuss the value of networking and community involvement in the data science field. Participating in events, conferences, and online forums can lead to fruitful collaborations and help you gain insights from diverse perspectives. We will provide tips on how to engage with the data science community and build lasting connections.

Lastly, we will emphasize the importance of continuous learning and skill development in the ever-changing landscape of data science. By staying curious and actively seeking new knowledge, you can deepen your understanding, enhance your skills, and open up new opportunities for career growth.

Throughout this module, we will provide practical examples and tips, ensuring that you have a clear roadmap to succeed in your data science career. So, let's get started on exploring the final thoughts and next steps in data science!

Building a Data Science Portfolio

A well-crafted data science portfolio is an essential tool for showcasing your skills, experience, and projects to potential employers and collaborators. In this section, we will guide you through the process of building a data science portfolio that effectively demonstrates your expertise and helps you stand out in a competitive job market.

Select your projects

Begin by choosing a diverse set of projects that demonstrate your proficiency in various data science techniques, such as data cleaning, visualization, machine learning, and statistical analysis. Make sure to include projects that highlight your ability to work with different types of data (e.g., structured, unstructured, time series) and various domains (e.g., finance, healthcare, marketing).

Document your projects

For each project, write a clear and concise description that outlines the problem you were trying to solve, the data you used, the methodologies you employed, and the results you achieved. Make sure to explain your thought process and decision-making rationale, as this will help potential employers understand how you approach data science challenges.

Include code samples

Including well-organized and commented code snippets is crucial, as it allows potential employers to assess your programming skills and see how you implement data science techniques. Make sure your code is readable and adheres to best practices, as this will reflect positively on your attention to detail and professionalism.

Showcase visualizations

High-quality visualizations can make your portfolio more engaging and help communicate complex insights effectively. Include examples of the visualizations you created during your projects, such as bar charts, scatter plots, heatmaps, and network graphs, and provide brief explanations of the insights they convey.

Write case studies or blog posts

In addition to project descriptions, consider writing more in-depth case studies or blog posts that delve into the details of your projects. This will allow you to showcase your communication and storytelling skills, which are essential for data scientists when presenting findings to non-technical stakeholders.

Highlight your domain expertise

If you have a background in a particular domain, such as finance or healthcare, make sure to highlight this in your portfolio. Showcasing your domain expertise can help you stand out to potential employers who are looking for data scientists with specialized knowledge in their field.

Include personal projects

Personal projects can demonstrate your passion for data science and your ability to learn and apply new concepts independently. Incorporate projects you have completed on your own or as part of online courses or hackathons, and discuss the motivation behind them.

Create a professional online presence

Choose a platform to host your portfolio, such as GitHub, GitLab, or a personal website. Ensure your online presence is polished and professional, with a consistent design and easy navigation. Include your contact information, a brief bio, and links to your LinkedIn profile and other relevant social media accounts.

Solicit feedback

Share your portfolio with mentors, peers, or experienced data scientists, and ask for constructive feedback. Their input can help you refine your presentation, identify gaps in your skill set, and improve the overall quality of your portfolio.

Keep your portfolio up-to-date

As you continue to learn and grow in your data science career, make sure to update your portfolio with new projects, accomplishments, and skills. Regularly updating your portfolio will demonstrate your commitment to continuous learning and professional development.

Quantify your impact

Whenever possible, include specific metrics or tangible outcomes in your project descriptions to demonstrate the impact of your work. For example, mention the percentage of improvement in a model's performance, the amount of revenue generated through your analysis, or the number of users impacted by your recommendations.

Highlight collaboration and teamwork

Data science projects often involve working with cross-functional teams. Be sure to mention instances where you collaborated with other team members, such as developers, product managers, or designers, to showcase your ability to work effectively in a team setting.

Showcase your unique strengths

Identify your unique selling points and make sure they are evident in your portfolio. For example, if you have a knack for creating innovative visualizations, make sure to feature them prominently. If you excel at natural language processing, include projects that demonstrate your expertise in this area.

Add testimonials or references

Including quotes from supervisors, colleagues, or clients who can vouch for your skills and work ethic can add credibility to your portfolio. Reach out to individuals who can provide a well-rounded perspective on your strengths and accomplishments.

Use multimedia

Consider incorporating multimedia elements, such as videos, interactive visualizations, or even podcasts, to make your portfolio more engaging and dynamic. This can also demonstrate your versatility and ability to communicate complex information in various formats.

Following these steps and being mindful of the details will allow you to create a comprehensive, polished data science portfolio that highlights your skills and achievements. A well-built portfolio will not only help you stand out in a competitive job market but also serve as a testament to your dedication and passion for the field of data science.

Staying Up-to-date with Industry Trends

Staying current with industry trends is crucial for your success as a data scientist, as it helps you maintain a competitive edge, adapt to new challenges, and identify emerging opportunities. In this section, we will discuss various strategies and resources for keeping up-to-date with the latest developments in the field of data science.

Follow reputable blogs and websites

Regularly read blogs and websites that cover data science news, tutorials, and best practices. Some popular options include Towards Data Science, KDNuggets, Data Science Central, and the DataCamp Blog. Make it a habit to set aside time each week to review new articles and learn from industry experts.

Subscribe to newsletters and magazines

Sign up for newsletters and magazines that focus on data science and related fields, such as machine learning, artificial intelligence, and big data. Some examples include the Data Elixir newsletter, the AI Weekly newsletter, and the O'Reilly Data Newsletter. These subscriptions will deliver curated content directly to your inbox, making it easy to stay informed.

Participate in online forums and communities

Engage in discussions and ask questions on data science forums, such as Reddit's r/datascience, Stack Overflow, and the AI section of arXiv. These platforms provide opportunities to learn from other professionals, share your own knowledge, and stay updated on current topics of interest.

Attend conferences and workshops

Participate in data science conferences and workshops, both in-person and online, to learn about the latest trends, network with professionals, and gain hands-on experience with new tools and techniques. Examples of popular data science conferences include NeurIPS, KDD, and the Data Science Bowl.

Take online courses and attend webinars

Continuously expand your skillset by enrolling in online courses and attending webinars. Platforms like Coursera, edX, and DataCamp offer a wide range of data science courses, from beginner to advanced levels. Webinars are another great way to learn from industry experts and stay informed about the latest developments.

Join professional organizations and networks

Become a member of professional data science organizations, such as the Association for Computational Linguistics (ACL), the Institute of Electrical and Electronics Engineers (IEEE), or the Association for the Advancement of Artificial Intelligence (AAAI). Membership in these organizations often provides access to exclusive resources, networking opportunities, and conferences.

Follow influential data scientists on social media

Keep an eye on the work and insights of leading data scientists by following them on platforms like Twitter, LinkedIn, and GitHub. This can help you stay updated on their latest projects, research, and thoughts on industry trends.

Monitor academic journals and preprints

Stay informed about cutting-edge research by regularly reviewing academic journals, such as the Journal of Machine Learning Research (JMLR), and preprint servers like arXiv.org. This will help you keep up with the latest developments in the field and gain a deeper understanding of new techniques and algorithms.

Collaborate on open-source projects

Contribute to open-source data science projects on platforms like GitHub or GitLab. This not only helps you stay current with new tools and techniques but also allows you to work with other data scientists and learn from their expertise.

Set up news alerts and RSS feeds

Use tools like Google Alerts or Feedly to set up news alerts and RSS feeds for specific data science keywords, such as "machine learning," "deep learning," or "data visualization." This will ensure that you receive regular updates on relevant news and research in the field, making it easier to stay informed about emerging trends and technologies.

Participate in hackathons and data science competitions

Join data science competitions and hackathons, such as those hosted on Kaggle, to challenge yourself and stay up-to-date with the latest problem-solving techniques. These events not only help sharpen your skills but also expose you to new ideas and perspectives from fellow data scientists.

Network with other data science professionals

Regularly engage with other data scientists through meetups, LinkedIn groups, or Slack channels. Building a strong network of peers can help you stay informed about industry trends, exchange ideas, and even discover new job opportunities.

Read books and whitepapers

Read books and whitepapers on data science topics to gain in-depth knowledge and stay current with the latest methodologies and best practices. Look for recent publications by leading authors and experts in the field to ensure you're learning from the most up-to-date resources.

Listen to podcasts

Subscribe to data science podcasts, such as Data Skeptic, DataFramed, or Not So Standard Deviations, to learn from industry experts and gain insights into current trends and challenges. Listening to podcasts is a convenient way to stay informed while commuting, exercising, or during your downtime.

Set personal learning goals

Regularly set and review personal learning goals to ensure that you are continuously expanding your knowledge and skillset. Identify specific areas of interest or gaps in your understanding, and develop a plan to address them through courses, reading, or hands-on projects.

Actively engaging in these strategies and utilizing a diverse range of resources will allow you to stay up-to-date with the latest industry trends and developments in data science. This will not only help you maintain a competitive edge in the job market but also enable you to adapt to new challenges and seize emerging opportunities in the rapidly evolving world of data science.

Networking and Community Involvement

Networking and community involvement play a crucial role in advancing your data science career, providing you with opportunities to learn from others, share your knowledge, and build valuable connections. In this section, we will discuss various strategies for engaging with the data science community and maximizing the benefits of networking.

Attend local meetups and events

Participate in local data science meetups, workshops, and events in your area. These gatherings provide an excellent opportunity to meet like-minded professionals, exchange ideas, and learn about the latest industry trends. Check out platforms like Meetup.com to find data science groups near you.

Join online communities

Engage in online data science communities, such as forums, Slack channels, or LinkedIn groups, where you can ask questions, share insights, and learn from other professionals. Some popular options include the AI section of arXiv, the Data Science Stack Exchange, and the Data Science Community on LinkedIn.

Attend conferences

Participate in data science conferences, such as NeurIPS, KDD, or the Data Science Bowl, to learn about the latest trends, research, and tools. Conferences are also a great place to network with fellow data scientists, industry experts, and potential employers.

Present your work

Share your knowledge and expertise by presenting your projects or research at conferences, meetups, or webinars. Public speaking can help you gain visibility within the data science community, demonstrate your communication skills, and open doors for new opportunities.

Collaborate on open-source projects

Contribute to open-source data science projects on platforms like GitHub or GitLab. Working on open-source projects allows you to collaborate with other data scientists, learn from their expertise, and showcase your skills to potential employers.

Participate in hackathons and competitions

Join data science hackathons and competitions, such as those hosted on Kaggle, to challenge yourself and connect with other data scientists who share similar interests. These events not only help you hone your skills but also provide an opportunity to collaborate, exchange ideas, and expand your professional network.

Build relationships with mentors and peers

Establish connections with experienced data scientists who can offer guidance, advice, and support throughout your career. Similarly, connect with peers who are at a similar stage in their career, as they can provide a valuable sounding board and offer insights based on their experiences.

Offer mentorship or guidance to others

As you gain experience and expertise in data science, consider mentoring or providing guidance to less experienced individuals. This not only helps you give back to the community but also strengthens your understanding of the subject matter and enhances your communication skills.

Engage on social media

Use social media platforms, such as Twitter or LinkedIn, to follow data science influencers, share your thoughts and insights, and engage in conversations with fellow professionals. Regularly engaging on social media can help you stay informed, build your online presence, and develop relationships with others in the industry.

Write articles or blog posts

Share your knowledge and insights by writing articles or blog posts on data science topics. Publishing your work on platforms like Medium or your personal blog can help you establish yourself as an expert in the field, attract the attention of potential employers, and connect with other professionals who share your interests.

Volunteer for data science projects

Offer your data science skills to support non-profit organizations, community initiatives, or open-source projects. Volunteering not only allows you to contribute to meaningful causes but also provides networking opportunities and helps you gain practical experience.

Leverage alumni networks

Connect with fellow alumni from your academic institutions or online courses, as they can offer valuable insights, job leads, or introductions to others in the data science field.

Set networking goals

Set specific, measurable networking goals, such as attending a certain number of meetups per month or connecting with a certain number of professionals on LinkedIn. Having clear goals will help you stay motivated and focused on building your network.

Follow up with connections

After making a new connection, be sure to follow up with a personalized message or email to express your appreciation for their time and reinforce your interest in staying in touch. Regularly engaging with your connections will help you build and maintain strong relationships within the data science community.

Actively participating in the data science community and leveraging various networking strategies will allow. You to build a robust professional network that will support your career growth and open up new opportunities. Networking and community involvement not only help you learn from others and stay current with industry trends but also enable you to make lasting connections that can enrich your professional journey.

Continuous Learning and Skill Development

In the rapidly evolving field of data science, continuous learning and skill development are essential for staying competitive and maintaining your expertise. In this section, we will explore various strategies and resources to help you stay current with new technologies, deepen your knowledge, and enhance your skillset.

Take online courses

Enroll in online courses to expand your knowledge and learn new skills. Platforms like Coursera, edX, and DataCamp offer a wide range of data science courses, from beginner to advanced levels. Dedicate time each week to complete coursework, participate in discussions, and apply your learnings to practical projects.

Attend webinars and workshops

Participate in webinars and workshops to learn from industry experts and gain hands-on experience with new tools and techniques. These events are a convenient way to stay informed about the latest developments and deepen your understanding of specific topics.

Read books and research papers

Regularly read books and research papers on data science topics to gain in-depth knowledge and stay current with the latest methodologies and best practices. Look for recent publications by leading authors and experts in the field to ensure you're learning from the most up-to-date resources.

Practice coding regularly

Improve your programming skills by coding regularly. Set aside time each week to work on coding exercises, personal projects, or contribute to open-source projects. Websites like LeetCode,

HackerRank, or Project Euler offer coding challenges to help you hone your skills and stay sharp.

Engage in personal projects

Develop and work on personal data science projects to apply your skills in real-world scenarios, explore new techniques, and deepen your understanding of the subject matter. Personal projects also provide an opportunity to showcase your skills in your portfolio.

Participate in data science competitions

Join data science competitions, such as those hosted on Kaggle or DrivenData, to challenge yourself and stay up-to-date with the latest problem-solving techniques. Competitions not only help sharpen your skills but also expose you to new ideas and perspectives from fellow data scientists.

Network with other professionals

Regularly engage with other data scientists through meetups, online forums, or social media. Networking can help you stay informed about industry trends, exchange ideas, and learn from the experiences of others.

Learn from blogs and podcasts

Follow reputable blogs and podcasts to stay current with industry news, tutorials, and best practices. This can help you learn about new technologies, tools, and techniques, as well as provide insights into the experiences and perspectives of other data scientists.

Set personal learning goals

Regularly set and review personal learning goals to ensure that you are continuously expanding your knowledge and skillset. Identify specific areas of interest or gaps in your understanding, and develop a plan to address them through courses, reading, or hands-on projects.

Obtain certifications

Pursue professional certifications to validate your skills and demonstrate your commitment to continuous learning. Some popular data science certifications include the AWS Certified Machine Learning - Specialty, Microsoft Certified: Azure AI Engineer Associate, and the Google Cloud Professional Data Engineer.

Attend conferences

Participate in data science conferences to learn about the latest trends, research, and tools. Conferences also provide opportunities to network with fellow data scientists, industry experts, and potential employers.

Teach or mentor others

Share your knowledge and expertise by teaching courses, leading workshops, or mentoring less experienced individuals. Teaching or mentoring not only helps you give back to the community but also strengthens your understanding of the subject matter and enhances your communication skills.

Learn new programming languages or tools

Stay versatile and adaptable by learning new programming languages or tools relevant to data science. Expanding your technical skills will help you tackle a broader range of projects and stay competitive in the job market.

Review and update your skillset

Periodically review your skillset and identify areas that may need improvement or updating. As new tools and techniques emerge, it's essential to ensure that your skills remain relevant and up-to-date.

Seek feedback from peers and mentors

Actively seek feedback from your peers, mentors, or supervisors to identify areas for improvement and gain insights into your strengths and weaknesses. Constructive feedback can help you create a targeted plan for continuous learning and skill development.

Stay informed about industry trends

Keep abreast of industry trends by following news, blogs, and social media accounts related to data science. Staying informed will help you anticipate changes in the field and adapt your learning plan accordingly.

Allocate time for learning

Set aside dedicated time each week for learning and skill development. Establishing a routine will help you maintain consistency and ensure that you're making steady progress in expanding your knowledge and abilities.

Embrace challenges

Step outside of your comfort zone and take on challenging projects or tasks that require you to learn new skills or apply existing ones in novel ways. Embracing challenges will accelerate your learning and help you grow as a data scientist.

Actively engaging in continuous learning and skill development will allow you to stay at the forefront of the data science field, adapt to new challenges, and ensure long-term career success. Developing a growth mindset and committing to lifelong learning will not only help you maintain a competitive edge in the job market but also enable you to seize emerging opportunities in the rapidly evolving world of data science.

Career Paths and Opportunities in Data Science

Data science offers a diverse array of career paths and opportunities, allowing you to apply your skills and knowledge in a variety of industries and roles. In this section, we will explore some of the most common and in-demand data science career paths, along with practical examples and advice to help you determine the best fit for your interests and goals.

Technical Career Paths

- **Data Analyst:** As a data analyst, you will focus on collecting, processing, and analyzing data to provide insights that inform business decisions. You will use statistical techniques, data visualization tools, and programming languages like Python or R to analyze data and create reports or dashboards. Data analysts often work in industries such as finance, healthcare, marketing, or retail.

- **Data Scientist:** Data scientists are responsible for designing and implementing machine learning models, conducting advanced statistical analysis, and creating data-driven solutions to complex problems. They work with large and complex datasets, using programming languages like Python or R, and machine learning frameworks like TensorFlow or PyTorch. Data scientists are employed across various industries, including technology, healthcare, finance, and e-commerce.

- **Machine Learning Engineer:** In this role, you will develop, deploy, and maintain machine learning models and systems. You will work with data scientists to implement their models and optimize them for production environments. Machine learning engineers often have strong programming skills in languages like Python or Java and experience with machine learning frameworks and cloud platforms.

- **Data Engineer:** Data engineers are responsible for building, maintaining, and optimizing the data infrastructure that supports data scientists and analysts. They design and implement data pipelines, ensuring that data is clean, reliable, and accessible. Data engineers often have strong programming skills in languages like Python, Java, or Scala, and experience with big data technologies like Hadoop or Spark.
- **Business Intelligence Analyst:** As a business intelligence analyst, you will focus on leveraging data to help organizations make data-driven decisions. You will create dashboards, reports, and visualizations to provide insights into business performance, trends, and opportunities. Business intelligence analysts typically have strong analytical skills, experience with BI tools like Tableau or Power BI, and a deep understanding of business processes.
- **Data Visualization Specialist:** In this role, you will specialize in designing and creating visually engaging and informative data visualizations. You will work with data analysts and data scientists to transform complex data into easily understandable visuals that effectively communicate insights to stakeholders. Data visualization specialists have expertise in tools like Tableau, D3.js, or ggplot2, and a strong understanding of design principles and storytelling techniques.
- **AI/ML Researcher:** As an AI/ML researcher, you will focus on advancing the field of artificial intelligence and machine learning by developing new algorithms, techniques, and approaches. You will conduct research, publish papers, and contribute to the scientific community. AI/ML

researchers typically have advanced degrees in computer science, mathematics, or a related field and strong programming skills.

- **Big Data Architect:** In this role, you will design and implement large-scale data processing systems and infrastructure to support data-driven initiatives. Big data architects have expertise in big data technologies like Hadoop, Spark, and NoSQL databases, as well as experience with cloud platforms like AWS, Azure, or Google Cloud Platform.

Management Career Paths

Data science professionals can also pursue various managerial career paths that involve overseeing teams, developing strategies, and driving data-driven decision-making within organizations. Here are some managerial career options for data science professionals:

- **Data Science Manager:** As a data science manager, you will lead a team of data scientists and analysts, overseeing their work and ensuring the delivery of high-quality data-driven solutions. You will also collaborate with other departments, manage project timelines, and contribute to strategic decision-making. Strong leadership, communication, and project management skills are essential in this role.

- **Analytics Manager:** In this role, you will manage a team of analysts responsible for extracting insights from data and creating reports, dashboards, and visualizations. You will ensure that your team's work aligns with business goals and helps drive informed decision-making. An analytics manager should have a solid understanding of analytics tools, business intelligence, and data visualization techniques.

- **Machine Learning Team Lead:** As a machine learning team lead, you will oversee a team of machine learning engineers and data scientists working on the development and deployment of machine learning models. You will be responsible for setting project goals, managing timelines, and ensuring the successful implementation of machine learning solutions. This role requires strong technical expertise in machine learning, as well as leadership and project management skills.

- **Data Engineering Manager:** In this position, you will manage a team of data engineers responsible for building and maintaining the data infrastructure that supports data-driven initiatives within the organization. You will oversee the design and implementation of data pipelines, data storage solutions, and data processing systems. A data engineering manager should have a strong background in big data technologies and programming languages.

- **Chief Data Officer (CDO):** As a CDO, you will be responsible for the overall data strategy, governance, and management within an organization. You will collaborate with other executives to ensure data is used effectively and ethically to drive business goals. This role requires a strong understanding of data management principles, data privacy regulations, and experience in creating and implementing data strategies.

- Data and Analytics Director: In this role, you will oversee the entire data and analytics function within an organization, including data scientists, analysts, and engineers. You will develop data-driven strategies, manage budgets, and ensure your team's work aligns with the organization's goals. A data and analytics director should have a deep understanding of data science, analytics, and business intelligence, along with strong leadership and communication skills.

To pursue a managerial career path in data science, you should focus on developing leadership skills, honing your communication abilities, and gaining experience in managing teams and projects. It's also essential to have a strong understanding of data science concepts, tools, and techniques, as well as a deep knowledge of the industry and business processes. Obtaining relevant certifications, attending workshops, and participating in leadership development programs can further enhance your qualifications for these managerial roles.

How to Choose the Best Career Path

To choose the best career path for you, consider your interests , strengths, and long-term goals. Here are some practical steps to help you identify the most suitable data science career path:

- **Assess your skills and interests:** Take an inventory of your current skills, knowledge, and interests. Reflect on the aspects of data science that you enjoy the most and identify areas where you excel. This will help you narrow down career paths that align with your strengths and passions.

- **Research different roles:** Conduct research on various data science roles to understand the specific responsibilities, skills, and tools associated with each position. This will give you a clearer idea of what each role entails and help you identify the best fit for your interests and skillset.

- **Attend industry events and webinars:** Participate in industry events, conferences, and webinars to gain insights into different data science roles and learn about emerging trends and opportunities. Networking with professionals in various roles can also provide valuable insights into what it's like to work in different positions.

- **Seek informational interviews:** Reach out to professionals working in data science roles that interest you, and request informational interviews. These informal conversations can provide invaluable insights into the day-to-day responsibilities, challenges, and rewards of different roles, helping you make a more informed decision about your career path.

- **Gain practical experience:** Work on personal projects, internships, or part-time roles that expose you to different aspects of data science. Gaining hands-on experience in various roles can help you better understand your preferences and identify the most suitable career path.
- **Pursue relevant education and training:** Once you've identified your desired career path, invest in the necessary education and training to develop the skills and knowledge required for that role. This may involve enrolling in online courses, obtaining certifications, or pursuing a degree program.
- **Network with professionals in your chosen field:** Actively engage with professionals working in your desired career path, as they can offer advice, insights, and potentially connect you with job opportunities. Attend industry meetups, join online forums, and participate in networking events to build relationships and learn from others in the field.

Thoroughly researching different data science career paths, assessing your skills and interests, and gaining practical experience, you will be able to make an informed decision about the best direction for your career. As you progress in your chosen path, continue to embrace continuous learning and skill development to stay competitive and adapt to the ever-changing landscape of data science.

Module 10 Exam Questions

1. What is the primary role of a data scientist?

a) Building data infrastructure

b) Designing visualizations

c) Developing machine learning models

d) Managing data storage solutions

Which programming language is commonly used in data science?

a) Python

b) C++

c) Swift

d) Ruby

Which library is used for numerical computing in Python?

a) Matplotlib

b) NumPy

c) Pandas

d) Seaborn

What is the primary purpose of descriptive statistics?

a) Predict future events

b) Summarize and describe data

c) Test hypotheses

d) Create machine learning models

Which of the following is a classification algorithm?

a) Linear Regression

b) Decision Trees

c) K-Means Clustering

d) Principal Component Analysis

What is the primary goal of unsupervised machine learning?

a) Predicting a continuous output

b) Predicting a categorical output

c) Identifying patterns and structure in data

d) Testing hypotheses

Which type of neural network is commonly used for image recognition tasks?

a) Convolutional Neural Networks

b) Recurrent Neural Networks

c) Feedforward Neural Networks

d) Radial Basis Function Networks

Which step in the data science process involves gathering and importing data from various sources?

a) Data Collection

b) Data Cleaning

c) Data Transformation

d) Data Visualization

What is the primary purpose of data visualization?

a) Storing data

b) Training machine learning models

c) Cleaning data

d) Communicating insights from data

Which skill is most important for a Business Intelligence Analyst?

a) Developing machine learning models

b) Designing and creating data visualizations

c) Building data pipelines

d) Researching new algorithms

What is the primary role of a data analyst?

a) Developing data strategies

b) Implementing machine learning models

c) Analyzing data and providing insights

d) Designing data storage solutions

Which tool is commonly used for creating interactive data visualizations?

a) TensorFlow

b) Tableau

c) Apache Spark

d) Scikit-learn

What is the main focus of inferential statistics?

a) Summarizing data

b) Predicting future events

c) Drawing conclusions from data samples

d) Cleaning data

Which of the following is a dimensionality reduction technique?

a) Logistic Regression

b) K-Nearest Neighbors

c) Principal Component Analysis

d) K-Means Clustering

What is the primary goal of supervised machine learning?

a) Clustering similar data points

b) Reducing the number of features

c) Learning from labeled data

d) Analyzing relationships between variables

Which library is used for data manipulation and analysis in Python?

a) Matplotlib

b) NumPy

c) Pandas

d) Seaborn

What is an essential skill for a data engineer?

a) Building data infrastructure

b) Designing visualizations

c) Developing machine learning models

d) Managing data storage solutions

Which of the following is a data preprocessing step?

a) Data Collection

b) Data Cleaning

c) Data Transformation

d) All of the above

Which technique is commonly used for exploratory data analysis?

a) Data Visualization

b) Dimensionality Reduction

c) Feature Engineering

d) Model Evaluation

What is a key component of a data science portfolio?

a) Personal projects

b) Job applications

c) Online courses

d) Industry conferences

Answers:

1. c
2. a
3. b
4. b
5. b
6. c
7. a
8. a
9. d
10. b
11. c
12. b
13. c
14. c
15. c
16. c
17. a
18. d
19. a
20. a

Conclusions and Next Steps

As we reach the end of this book, it is evident that the Data Science revolution is transforming the way we make sense of the world around us. The impact of data-driven decision-making on our personal and professional lives is profound. Throughout this journey, we have examined the power of data to drive innovation, revealed its potential to enhance human potential, and contemplated the ethical and societal implications of data-driven insights. We have also delved into strategies for fostering a data-driven mindset, navigating the ethical landscape, and building data leadership and organizational readiness.

The future of work will be shaped by the seamless integration of Data Science with human intuition, critical thinking, and adaptability. By understanding the strengths and limitations of Data Science, we can leverage its potential to create a sustainable and equitable future for all. However, to make a genuine difference, it is crucial for each individual to take proactive steps towards embracing the changes brought by the Data Science era.

As a game changer, you have the power to make a lasting impact on your career and the world. Here are the next steps you can take to embrace the Data Science revolution and create positive change:

1. **Enroll the online program at TheBestMBA.org:** Certify your skills, gain in-depth knowledge, practical tools, and valuable insights from our comprehensive curriculum. Our program is designed to help you thrive in the Data Science era and adapt to the ever-changing demands of the workforce.
2. **Engage in continuous learning:** Stay ahead of the curve by constantly updating your skills and knowledge. Embrace lifelong learning and take advantage of the vast resources available online, such as courses, webinars, podcasts, and articles.
3. **Expand your network:** Connect with industry leaders, experts, and fellow game changers through Data Science

focused events, conferences, and workshops. Foster meaningful relationships and collaborations that will propel you forward in your career.

4. **Seek mentorship and guidance:** Reach out to experienced professionals in the field of Data Science and related industries to learn from their insights and experiences. Their wisdom can help you navigate the complexities of the data-driven landscape and ensure your success.
5. **Volunteer or collaborate on Data Science projects:** Gain hands-on experience and demonstrate your capabilities by working on real-world Data Science projects. This not only enhances your skills but also showcases your value to potential employers and collaborators.

By taking these steps, you will be well on your way to becoming an influential game changer in the Data Science era. Remember, the journey towards a sustainable and inclusive future begins with each individual's commitment to adapt, learn, and grow. Embrace the possibilities that Data Science offers and let your unique human skills shine through as you create a better tomorrow.

The future is now. Are you ready to shape it?

Join the IIEC Data Science Certification Program today available at https://thebestmba.org and become a game changer.

I hope you enjoyed this journey together, and as I mentioned at the beginning of this book, I would love for you to share your experience with me after completing this program. Feel free to message me via LinkedIn.

https://www.linkedin.com/in/angelberniz

To your success,

Angel Berniz

Made in the USA
Monee, IL
04 November 2026

67384048R00203